The Cardiovascular System During Exercise and Recovery

Tatsuhisa Takahashi

The Cardiovascular System During Exercise and Recovery

Springer

Tatsuhisa Takahashi
Department of Mathematical Information Science
Asahikawa Medical University
Asahikawa, Hokkaido, Japan

ISBN 978-981-99-8485-5 ISBN 978-981-99-8486-2 (eBook)
https://doi.org/10.1007/978-981-99-8486-2

This Springer imprint is published by the registered company Springer Nature Singapore Pte Ltd.
The registered company address is: 152 Beach Road, #21-01/04 Gateway East, Singapore 189721, Singapore

Preface

Many books have been published on the circulatory dynamics and cardiovascular control during exercise so far. Even now, such books are probably being written somewhere in the world. It goes without saying that the contents of these books written by talented authors are scientifically valuable. In addition, the continuous development of research is largely due to the fact that various studies by many researchers have been carried out and the results have been left as literature all over the world; it is regrettable that the important contributions of all authors are not specifically listed in this book.

In general, most of the research on cardiovascular control in exercise and physical activity is mainly at the beginning of exercise or during exercise. Conversely, there are relatively fewer studies of circulatory control during the periods of postexercise and recovery from exercise. Further, most of the studies have shown an interest in blood flow and pressure as measured at rest and during exercise in a standing or a sitting position. Many papers give the impression that observations of circulatory control at rest are comparable in standing, sitting, and supine positions. For exercise, however, research reports provide more information for the standing and sitting positions, as against the supine position. This is because the standing position is the posture when walking or running, and the sitting position is the posture in which cycling exercise is performed. In particular, the number of studies on cardiovascular recovery responses from exercise is relatively small, and there are extremely few studies that compare them in the standing and sitting positions with those in the supine position.

As yet, a large number of experiments have been conducted in quadrupedal animals; however, the distribution of blood flow to the body in quadrupeds is significantly different from that in bipedal humans. Therefore, the data and findings obtained from quadrupedal animals often limit our understanding of the blood distribution in the humans correctly.

In daily life, we often use our body for various work and activities, besides exercising for health. After these physical activities, there is always a period of rest, that is, a time period corresponding to a recovery period. It is the fact that a circulatory change during the recovery period after exercise does not completely represent the reversal against that at the start of exercise. Thus, the mechanisms responsible for circulatory control during the recovery period after exercise, in which we have been greatly interested, have not been fully elucidated. In particular, there are still unexplained problems regarding the differences in cardiovascular control between the upright and supine positions during the recovery period after exercise.

This book is composed of seven chapters. The first chapter introduces the enhanced recovery of heart rate to its preexercise resting level by light exercise, compared with heart rate during complete-rest recovery. The second chapter deals with similar time courses of mean blood pressure during recovery from exercise in an upright and a supine position. The similar responses of blood pressure are achieved through the cardiovascular regulation of heart rate, stroke volume, cardiac output, and peripheral vascular resistance in a reciprocal manner by offsetting each other between the two positions. The third chapter presents that the recovery of slowly decreasing heart rate after exercise is facilitated by cool-down exercise. This facilitated recovery of heart rate is dependent on the duration of cool-down exercise and is associated with an increase in cardiac efferent vagal tone. The fourth chapter addresses that a biphasic change in heart rate, or a sharp increase and a subsequent decrease, at the onset of light exercise occurs from rest in upright, but not in supine posture. The complex changes in heart rate are discussed with circumstantial evidence that withdrawal of vagal nerve activity is mediated by central command and subsequent vagal activation is mediated by cardiopulmonary mechanoreceptor reflexes due to augmented venous return of blood to the heart by the skeletal muscle pump. The fifth chapter treats postexercise regulation of the cardiovascular system between inactive and active recovery from moderate cycle exercise in an upright and a supine position. The discussion section addresses that the integrated interplay between the heart and vasculature enables the cardiovascular system to control arterial blood pressure. The sixth chapter deals with the differences in blood flow velocity in the femoral artery measured by Doppler ultrasound velocimetry between inactive, passive, and active recovery. The discussion section explains a role of wall shear stress of blood flow in the femoral artery and the total number of capillaries in the skeletal muscle tissues. The seventh and final chapter is concerned with the Doppler measurement of blood flow velocity in the middle cerebral artery at rest and during cycle exercise and a comparison of measurements between men and women. Numerical calculation also demonstrates that differences in blood velocity and hematocrit between men and women are related to efficient oxygen transport to the middle cerebral artery and the dependent tissues.

This book has been written for a broad audience that includes medical students and residents, graduate students in the medical sciences, kinesiology, biomedical engineering, and sport medicine, specialists in aerospace medicine and gravitational physiology, cardiologists, and any physician or medical professional with an interest in human cardiovascular function.

The late Dr. Akira Kamiya, my supervisor, professor of biomedical engineering at the University of Tokyo, once told me "water and oxygen are the most important elements for all living creatures, not just mammals." He also told that the maintenance of tissue water and oxygen is in accordance with the design principle for the structure and function of the respiratory and circulatory systems. When systemic arterial blood pressure is maintained normally, the body is structured so that blood can flow optimally through parts and all of the circulatory system and then the supply of water and oxygen to tissues can be maintained normally. For this reason, systemic arterial pressure is the primary variable regulated by the various mechanisms of baroreceptors, mechanoreceptors, chemoreceptors, feedback systems, catecholamines, hormones, and the autonomic nervous system. In addition, shear stress acting on blood vessels is now well known for its important role in regulating blood flow. The function of shear stress was probably first introduced in an elaborate study in animals by Kamiya. In this book as well, we examine how the cardiovascular variables, such as heart rate, stroke volume, cardiac output, and total peripheral resistance, work together. Then, it demonstrates that arterial pressure is regulated during postural changes and postexercise recovery. The number of capillaries in the skeletal muscle tissue is also estimated in Chap. 6. The range of tissues that the capillaries nourish is described, what it means is that they have an optimal distance for oxygen diffusion.

After writing this book, I have realized that the research I have done thus far has been guided by what I learned from a great physiologist, Kamiya, when I was aspiring to be a physiologist.

Asahikawa, Hokkaido, Japan Tatsuhisa Takahashi

Contents

About the Author

Tatsuhisa Takahashi, Ph.D. Professor, Asahikawa Medical University.

1992: Special Research Student at Graduate School of Medicine, The University of Tokyo.

1993: Graduate School of Engineering, Department of Bioengineering, Hokkaido University.

1993: Research Assistant, Department of Biomedical Engineering, Yamagata University.

2002: Associated Professor, Department of Mathematical Information Science, Asahikawa Medical University.

2021: Professor, Department of Mathematical Information Science, Asahikawa Medical University.

Specialty: Fractal physiology, Microcirculation, Exercise physiology, Circulatory physiology, Respiratory physiology, and Visual cognitive science.

Education: Information literacy, Statistics, Data science.

Books:

(1) Microcirculation in Fractal Branching Networks. Springer, Tokyo, 2014

ISBN 978-4-431-54507-1

eBook ISBN 978-4-431-54508-8

(2) Easy to Understand Data Analysis and Statistics: Let's Start Analysis Statistics of Medical and Life Science with Excel. Ohmsha, Ltd., Tokyo, 2018

POD ISBN 978-4-274-70061-3

eBook ISBN 978-4-274-80369-7

(3) The World of Cognitive Science: Why Can People Read Fragmented Letters? Ohmsha, Ltd., Tokyo, 2019

POD ISBN 978-4-274-70099-6

eBook ISBN 978-4-274-80407-6

(4) わかりやすいデータ解析と統計学:医療系の解析統計をExcelではじめてみよう. オーム社, 2017

ISBN 978-4-274-22111-8

Chapter 1
Cardiac Responses to Inactive and Active Recovery

1.1 Introduction

When we perform muscular exercises, the active muscles are supplied with increased oxygen, nutrients, and chemical substances by the circulation. These essential substances are sufficiently transported with an increase in blood flow and are regulated by the cardiovascular system. The increase in muscle blood flow is due to an increase in cardiac output that is accompanied with an increase in venous return of blood to the heart. The increase in muscle blood flow is also associated with the redistribution of a large volume of blood from inactive muscles and splanchnic regions to active muscles during exercise.

An increase in cardiac output is produced by an increase in left-ventricular stroke volume (the volume of blood ejected per left ventricular contraction) and an increase in heart rate; thus, the cardiac output is defined as the product of stroke volume and heart rate. The increase in cardiac output during exercise is of importance for maintaining the homeostasis in the muscular tissues. Oxygen and nutrition are transported by the blood to the working muscle tissues. At the same time, the metabolic by-products, such as carbon dioxide, lactate, and heat, generated in the muscles are carried away from the tissues by the blood. The circulatory system also coordinates with the respiratory system that plays an important role in exchanging gases, oxygen, and carbon dioxide between air and blood in the alveoli.

Heart rate and oxygen uptake increase rapidly at the onset of exercise and then increase gradually to their steady levels during submaximal exercise at a constant intensity (Fig. 1.1). On the other hand, heart rate and oxygen uptake increase linearly as exercise intensity increases progressively. The amount of oxygen that is transported to produce energy in the mitochondria in the muscle tissues is proportional to the exercise intensity. The relationship between changes in heart rate and oxygen uptake indicates a simple regression line from rest to submaximal exercise.

T. Takahashi, *The Cardiovascular System During Exercise and Recovery*,
https://doi.org/10.1007/978-981-99-8486-2_1

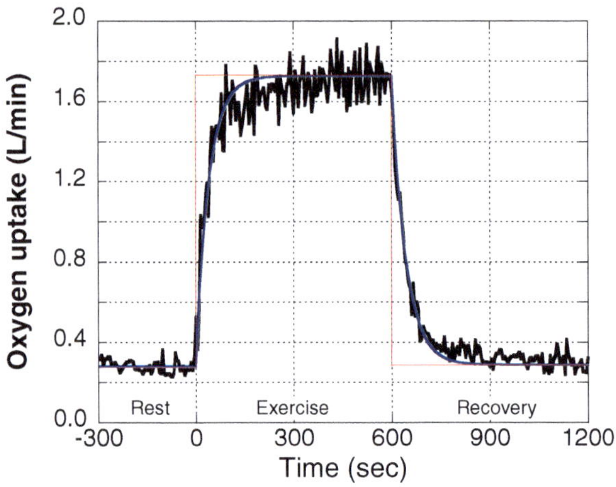

Fig. 1.1 Breath-by-breath oxygen uptake responses to heavy exercise at a step increase and a step decrease in work load on the basal resting level in a single subject. The solid line represents a best-fit exponential function to the measurements. The kinetics of oxygen uptake is quantified by time constants of 41.0 s for the on-transient and 43.0 s for the off-transient response. The data points for oxygen uptake were subjected to least-squares fitting with an approximate equation as a function of time: (1) $y = \text{baseline} + \Delta \cdot [1 - \text{Exp}(-x/\text{TC})]$ for on-response; (2) $y = \text{baseline} + \Delta \cdot [\text{Exp}(-x/\text{TC})]$ for off-response, where Δ is the difference between baseline and steady-state values and TC is the so-called "time constant" used in control engineering or electrical engineering to quantify the responsive property of a fitted curve. The method of fitting a curve to oxygen kinetics is detailed in the author's book, Easy to Understand Data Analysis and Statistics. Referring to the book, you can easily calculate time constants with Excel

During recovery after the end of an exercise with a higher work load, heart rate exponentially decreases to a relatively constant level; however, the decreased heart rate remains higher above a baseline level at rest before exercise. This higher recovery heart rate is sustained for a long time and far beyond a time during which the respiratory and neuroendocrine systems gradually return to their resting levels. It takes a relatively long time to restore the respiratory and circulatory systems that remove metabolites from the active tissues and then the whole body. Both the systems are also responsible for the delivery of oxygen and nutrients in the opposite manner. The time length of recovery heart rate to the pre-exercise resting level is dependent on both the intensity and duration of the exercise; that is, the stronger or longer the muscles exercise, the longer it takes to regain one's heart rate.

So far, the mechanisms responsible for the slower recovery of heart rate have not been well understood. Moreover, it was unclear whether this recovery of heart rate could be facilitated by a cool-down exercise during recovery (active recovery) until Takahashi and his coworkers undertook a study of that theme. In contrast, it has been well known that adequate exercise for recovery makes it possible to hasten the removal of metabolites, for example, lactate, from the skeletal muscles and increase the circulation effectively with higher blood flow to the muscles, as compared with complete rest during recovery (inactive recovery). In those who are not athletes, the plasma concentration of lactic acid progressively increases as the exercise intensity becomes greater than a relative intensity of about 50% of maximal oxygen consumption; accordingly, an appropriate intensity of cool-down exercise is likely to depend on the fitness level of individuals. The removal rate of lactate from the circulation is higher during mild or moderate exercise than during inactive recovery of complete rest. Thus, appropriate exercise during the phase of recovery from exercise can play an important role in facilitating restoration from the metabolic perturbations. In contrast, very little is known about whether continuous exercise during recovery from intensive exercise affects the restoration of the cardiovascular system or how light or mild exercise affects the cardiac responses during the recovery phase, in particular.

As a focus, this chapter addresses how the time courses of heart rate, stroke volume, and cardiac output during recovery from a heavy cycle exercise are affected by a cool-down exercise at a light intensity. We evaluate the hypothesis that the cool-down exercise during recovery from exercise promotes the recovery of the circulatory system to the baseline level at rest before exercise. In this study, after a 10-min bout of exercise at a constant work rate of 170 watts, metabolic (oxygen uptake) and cardiac variables are monitored during complete rest (inactive recovery) and during light exercise at 20 or 40 watts (partially active recovery), each one of which is followed by complete rest. The increased cardiorespiratory variables that accompany continuous limb movements during active recovery would lead to the subsequent decline of these variables toward resting levels after active recovery.

1.2 Materials and Methods

1.2.1 Human Subjects

Five healthy young male college students participated in this study. The physical characteristics of the volunteers with a mean age of 22.2 ± 0.8 (mean ± SD) years were as follows: a mean height of 1.69 ± 0.05 m, a mean body mass of 59.2 ± 1.6 kg, and a mean peak oxygen uptake of 2.55 ± 0.26 L/min. All the participants were fully informed of the procedures, risks, and benefits of the study after a medical examination. None of them had any history of circulatory or respiratory disease. Before the experiment, a written informed consent was obtained from all the subjects, who were thoroughly acquainted with the experimental procedure. This research and experiment were conducted with the approval of the Health Research Ethics Board at the university. This study was also carried out in compliance with the principles outlined in the Declaration of Helsinki regarding experimentation with human participants. The participants complied with the instructions that they were not involved in any regular physical activity or competitive sports.

1.2.2 Cardiac Measurements

Heart rate (HR, in beats per minute), stroke volume (*SV*, in milliliters), and cardiac output (*CO*, in liters per minute) were recorded with a computerized measurement system. R-R intervals (the times between consecutive heartbeats) were measured with an electrocardiograph (ECG) through three bipolar leads fastened to the chest. Heart rate was calculated as the reciprocal of R-R intervals.

An impedance cardiograph (ICG) with four electrodes (RGA-5; Nihon Koden, Tokyo, Japan) was used to monitor bioelectric, transthoracic impedance for determining stroke volume noninvasively. For measurement of the bioimpedance, the two circumferential strip electrodes (leads 1 and 2), 5-mm width and 0.5-mm thickness, of copper were placed 30 mm apart at the neck and the other two (leads 3 and 4) were placed at the abdomen, the outline of which is shown in Fig. 1.2. It is best to place the electrode 2 around the base of the neck and the electrode 3 around the level of the xiphisternal joint. A 100-kilohertz (kHz), 350-microampere (μA) sinusoidal current was fed into the two outside electrodes 1 and 4. The electrical impedance between the two inside electrodes 2 and 3 was continuously recorded.

Stroke volume was calculated from an empirical equation derived by Kubicek et al.:

$$\mathrm{SV} = \rho\left(L / Z_0\right)^2 \left(dZ / dt\right)_{\mathrm{min}} T_{\mathrm{ET}} \tag{1.1}$$

where ρ (ohm·cm) was the electrical resistivity of blood, L (cm) was the mean distance between the inner potential electrodes by averaging the anterior and posterior distances, Z_0 (ohm) was the baseline impedance across the two inner electrodes, $(dZ/dt)_{\mathrm{min}}$ (ohm/s) was the minimum value of the first derivative of the change in thoracic impedance (ΔZ) with respect to time, and T_{ET} (s) was the ventricular ejection time obtained from the dZ/dt waveform. The relationships among these

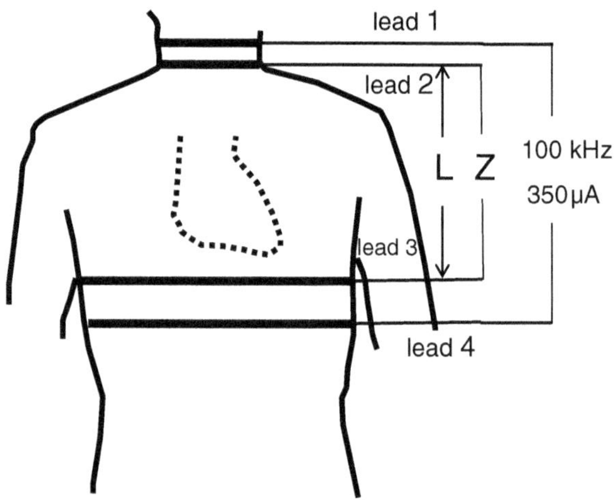

Fig. 1.2 Typical positions of four circumferential impedance electrodes. A constant sinusoidal alternating current is introduced through leads 1 and 4. Leads 2 and 3 detect changes in voltage. L stands for the distance between inner receiving electrodes. ΔZ is the variation of impedance signal

parameters in terms of impedance waveforms are shown in Fig. 1.3 with simultaneous phonocardiogram and electrocardiogram taken in a subject.

T_{ET} was measured in time from the last zero-line crossing of dZ/dt before the $(dZ/dt)_{min}$ to the first positive peak following the $(dZ/dt)_{min}$. Theoretically, the starting point for the determination of T_{ET} was a representative better than the use of 15% of the $(dZ/dt)_{min}$; however, Kubicek et al. have proposed that $0.15 \cdot (dZ/dt)_{min}$ was also a reliable point for determining T_{ET}, because it could avoid a difficulty in making the start point of T_{ET} from a small oscillation of the waves.

The electrical resistivity of blood can be estimated with hematocrit (*Hct*, in percent) from the formula:

$$\rho = 66(3 + 0.019Hct)/(3 - 0.038Hct) \tag{1.2}$$

where *Hct* is the percentage of volume of red blood cells occupying a blood sample.

The ECG and ICG electrodes attached to the surface of the body did not affect spontaneous breathing at rest and during exercise. Some noises of impedance signals caused by respiration and body motion were eliminated by two mathematical techniques: (1) moving average was the average of ICG waveforms over 10 cardiac cycles; and (2) ensemble average was the average of two measurements obtained from two trials of the same exercise test in each individual.

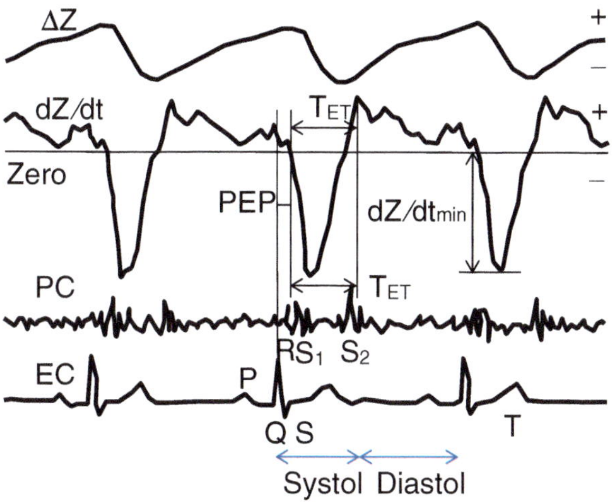

Fig. 1.3 A set of impedance-cardiography recordings in a normal subject: ΔZ, the variation of impedance signal; dZ/dt, the first derivative of ΔZ; dZ/dt_{min} the minimum dZ/dt, *PEP* pre-ejection period, T_{ET} left ventricular ejection time, *PCG* phonocardiogram, *ECG* electrocardiogram. A decrease in impedance is recorded as a downward deflection in ΔZ and dZ/dt tracings. Note that the QRS complex (depolarization of the ventricles) occurs at the beginning of systole and the rise in ventricular pressure. The *T* wave (repolarization of the ventricles) occurs at the same time that the ventricles relax at the beginning of diastole

Although absolute values for stroke volume derived from impedance signal remain slightly unsettled, there has been a general agreement that ICG measurement is reliable for estimating relative changes in stroke volume from rest to exercise and recovery. We checked daily variation in intraindividual estimates for stroke volume and attempted to place the electrodes of ICG at the same positions. The reproducibility of electrode positions is crucial for the measurement of stroke volume because the estimates are a function of the distance between the electrodes, as shown in Eq. (1.1).

1.2.3 Respiratory Measurements

Tidal volume (V_T, in liters per breath), breathing frequency (f, in breaths per minute), pulmonary expired ventilation (V_E, in liters per minute), oxygen uptake (VO_2, in liters per minute), and carbon dioxide output (VCO_2, in liters per minute) were measured breath-by-breath with a computerized measurement system. Subjects wore a face mask, the front entrance of which had a thermal wire transducer. Inspiratory and expiratory gases can flow bidirectionally through the thermal wire transducer. By this transducer, the bidirectional flow was converted into the electrical signal that was transmitted to a pneumotachograph (RF-2; Minato, Osaka, Japan); thus, air taken into the lungs and then taken out from the mouth were monitored as ventilatory airflow with this pneumotachograph. Small samples of gases through an outlet pipe in the side wall of the face mask were introduced into a medical mass spectrometer (WSMR-1400; Westron, Chiba, Japan) via a narrow-bore tube of only a little dead-space volume. Exhaled gas composition was continuously analyzed with the medical mass spectrometer. The spectrometer was calibrated with a standard reference gas mixture before each measurement.

The expired volume per minute (V_E) and the amount of gas moved per breath (V_T) were corrected to BTPS: body temperature, ambient pressure, and saturated with water vapor. Oxygen uptake and carbon dioxide output were expressed in STPD: standard temperature (0 °C), standard pressure (760 mmHg), and dry (no water vapor).

1.2.4 Preliminary Test

A week before the first exercise trial, each subject underwent a ramp exercise test with an incremental work rate of 15 watts every minute for determination of the peak of oxygen uptake, abbreviated to VO_{2peak}. The exercise was conducted on a stationary cycle ergometer with an electromagnetic brake resistance (Corival 300; Lode, The Netherlands) until the subject could not continue to exercise any more. This end point termed "exhaustion" was taken as a point when the subject was unable to maintain a pedaling rate of at least 50 revolutions per minute. The peak oxygen uptake was defined as the highest oxygen uptake obtained from the mean value over a 10-s period of breath-by-breath measurements; by definition, the term peak oxygen uptake is used when the test performance appears to be limited by local muscular factors rather than central circulatory dynamics.

1.2.5 Experimental Protocols

All subjects underwent the same exercise and then three kinds of recovery as shown in Fig. 1.4; they conducted each exercise testing twice. The subjects rested for 5 min in an upright seated position on the cycle ergometer in order to obtain pre-exercise cardiorespiratory basal levels and then exercised at a work rate of 170 watts for 10 min. After exercise, they took recovery for a total of 10 min on the cycle ergometer under three different conditions: (1) inactive recovery, complete rest for 10 min; (2) partially active recovery I, cycling exercise at a work rate of 20 watts for 7 min followed by quiet rest for 3 min; and (3) partially active recovery II, cycling exercise at a work rate of 40 watts for 7 min followed by quiet rest for 3 min. All the subjects had no particular differences in physical characteristics or peak oxygen uptake. The same absolute intensity was then set for the stimulus exercise and two active-recovery exercises they performed.

Two steady-state exercises at constant workloads of 20 and 40 watts for 5 min were each performed without the preceding intensive exercise to determine the relative intensity of active-recovery exercise as a percentage of peak oxygen uptake for each subject.

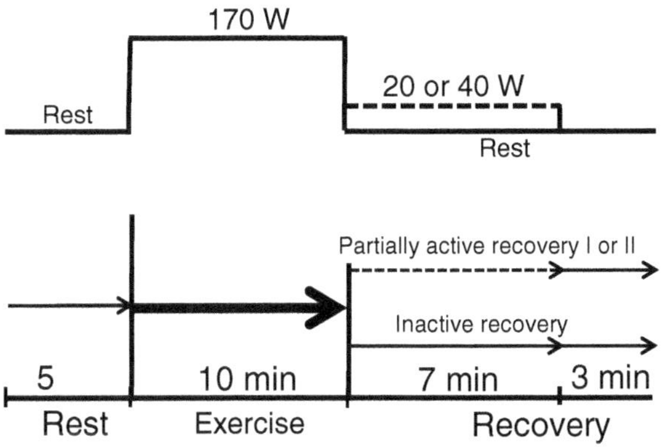

Fig. 1.4 A schematic diagram of exercise protocols. Exercise testing consisted of rest for 5 min, cycle exercise for 10 min at a work rate of 68% of peak oxygen uptake, and recovery for 10 min. The three modes of recovery performed separately were as follows: (1) complete rest (inactive recovery) for 10 min in a sitting position on the ergometer, (2) light pedaling at 20 watts for 7 min followed by sitting rest for 3 min (partially active recovery I), and (3) light pedaling at 40 watts for 7 min followed by sitting rest for 3 min (partially active recovery II)

The subjects pedaled at a constant rate of 60 revolutions per minute and maintained the pace with the aid of an audible metronome throughout the experiments. The pedaling rate was checked with an inductive revolution meter fitted to the ergometer. They were instructed to place their feet on the footrest next to the flywheel during the rest periods, thereby minimizing the facilitated venous return of blood to the heart by the muscular pumping action associated with movement of the lower extremities.

1.2.6 Experimental Schedule

All the tests were conducted between 10:00 and 17:00. The time of days at which an exercise test was conducted varied from subject to subject. However, each subject performed the trials at almost the same time of the days. The unified time suppressed the influence of a circadian rhythm on a long-term change in the autonomic nervous system. The order of tests was randomly assigned to avoid an order effect among the subjects. Each subject completed no more than one of three different tests on a given day. Each performed the same two trials with a separation of at least 3 h so that all cardiorespiratory variables could completely return to pre-exercise resting levels before the second trial. The subjects were scheduled to participate in the experiments 1 or 2 days a week, thus restraining exercise training from affecting the cardiorespiratory responses.

The subjects were instructed to abstain from any beverages containing caffeine or alcohol for at least 12 h before the exercise tests. They were allowed to have a light meal between 2 and 4 h before the tests so that the effect of the meal on the cardiovascular system could be excluded.

The subjects wore light clothing and sports shoes. All experiments were conducted at a room temperature of 19–23 °C (degrees Celsius).

1.2.7 Data Analysis

Lagrange polynomial formulas were fitted to both breath-by-breath and beat-by-beat data recorded in every test, specifically, by Lagrange interpolation, where the formulas estimate the values of intermediate data at regular intervals of 5 s. The interpolated data obtained from two repetitions of each test were averaged to yield a single set of data for each subject. Group mean values were then obtained from the mean sets of data for five subjects. Consecutive 10-s averages for all variables were computed to compare changes in time course between three kinds of recovery. The steady-state mean values for all variables were determined from the last minute at rest before exercise and during exercise. Two mean values for all variables were also determined at the seventh and tenth minutes during recovery. The former was from the last minute of the active recovery period and the latter was from the last minute of 3-min rest after active recovery.

1.2.8 Statistical Analysis

Statistical differences in mean values for oxygen uptake, heart rate, stroke volume, and cardiac output before, during, and after exercise were analyzed by using one-way analysis of variance (ANOVA). Analysis of variance was also used to compare the mean values for all variables at the tenth minute of three different modes of recovery with those at rest before exercise; this analysis confirmed whether these variables returned to resting baseline levels. If the analysis found a statistical significance in the trend of mean values, pairwise comparison was performed by using the Scheffé's test. Differences were considered significant for all statistical analyses at $P < 0.05$. Data are presented as mean ± standard deviation (SD).

1.3 Results

Table 1.1 summarizes the mean values for all variables obtained from five subjects at rest before exercise, during heavy exercise at a relative intensity of 68 ± 4% of peak oxygen uptake, and during one inactive and two partially active recoveries. There were no significant differences in mean value either at rest or during the same exercise before the three different recovery protocols. Metabolic and cardiac variables were significantly ($P < 0.05$) increased from rest to exercise; however, stroke volume increased only slightly.

Figure 1.5 shows the time courses of oxygen uptake during inactive recovery of complete rest on the ergometer and during two partially active recoveries with light-load pedaling at 20 and 40 watts and subsequent rest on the ergometer. The absolute intensities of 20 and 40 watts were equivalent to relative intensities of 25 ± 2% and 28 ± 3%, respectively, of peak oxygen uptake. Oxygen uptake during inactive recovery and two active recoveries decreased rapidly during the first 2 min and declined slowly thereafter. The mean oxygen uptake at the seventh minute was significantly greater ($P < 0.05$) for two active recoveries than for inactive recovery (Table 1.1); however, the difference between the two active recoveries was not significant. In each of the three recovery experiments, the restoration of oxygen uptake from exercise to the pre-exercise resting level was fully completed within 10 min of recovery.

Figure 1.6a shows the time courses of heart rate obtained from five subjects during inactive recovery and two partially active recoveries with light-load pedaling. Heart rate during inactive recovery exponentially decreased to a relatively stable level within almost 6 min. The mean heart rate at the tenth minute of inactive recovery was still significantly ($P < 0.05$) greater by about 14 beats/min than that at rest before exercise. In contrast to the difference in oxygen uptake response, the time courses of heart rate during two active-recovery exercises were nearly identical to that during inactive recovery. After the end of two recovery exercises, the two levels of resting heart rate further decreased separately and reached each of their constant levels; however, the difference was not significant ($P = 0.539$) in the mean heart rate at the tenth minute. The heart rates at the third minute after active recovery of 7 min

Table 1.1 Cardiovascular variables before, during, and after exercise

Inactive recovery experiment					
			Recovery		**ΔHR (beats/min)**
	Pre-exercise rest	**Exercise**	**Rest**	**Rest**	
VO_2 (L/min)	0.29 ± 0.05	1.71 ± 0.13	0.30 ± 0.05	0.28 ± 0.05	
HR (beats/min)	74.0 ± 7.9	134.1 ± 7.6	92.8 ± 7.9	91.4 ± 5.9*	1.4 ± 2.5
SV (mL)	92.8 ± 8.1	94.8 ± 14.0	80.7 ± 14.6	78.7 ± 14.5*	
CO (L/min)	6.8 ± 0.6	12.7 ± 1.8	7.5 ± 1.5	7.2 ± 1.4	
Partially active recovery experiment I					
			Recovery		
	Pre-exercise rest	**Exercise**	**20 W**	**Rest**	**ΔHR (beats/min)**
VO_2 (L/min)	0.33 ± 0.04	1.73 ± 0.07	0.62 ± 0.09†	0.31 ± 0.02	
HR (beats/min)	76.0 ± 8.4	133.4 ± 9.4	93.5 ± 8.7	86.0 ± 6.2	7.5 ± 4.4†
SV (mL)	93.9 ± 10.1	98.3 ± 11.2	101.3 ± 3.6†	86.6 ± 11.8	
CO (L/min)	7.1 ± 0.7	13.1 ± 1.6	9.5 ± 1.0†	7.5 ± 1.2	
Partially active recovery experiment II					
			Recovery		
	Pre-exercise rest	**Exercise**	**40 W**	**Rest**	**ΔHR (beats/min)**
VO_2 (L/min)	0.31 ± 0.04	1.71 ± 0.11	0.70 ± 0.12†	0.29 ± 0.05	
HR (beats/min)	74.8 ± 8.5	131.2 ± 8.0	94.5 ± 6.2	84.4 ± 7.4	10.0 ± 3.1†
SV (mL)	90.6 ± 14.3	96.6 ± 12.0	101.3 ± 5.1†	86.7 ± 13.7	
CO (L/min)	6.7 ± 0.7	12.6 ± 1.4	9.6 ± 0.8†	7.3 ± 1.0	

Values are means ± SD for five subjects undergoing each of the three experiments
VO_2 oxygen uptake, *HR* heart rate, *SV* stroke volume, *CO* cardiac output, *20 W* cycling at 20 watts, *40 W* cycling at 40 watts, Δ*HR* reduction in mean value from the seventh to tenth minute
**P* < 0.05, significantly different from the pre-exercise resting value; †*P* < 0.05, significantly different from the corresponding value for inactive recovery
Adapted from Takahashi and Miyamoto (1998), reproduced by permission

were both significantly ($P < 0.05$) lower than that during inactive recovery at the tenth minute: 91.4 ± 5.9 beats/min for inactive recovery, 86.0 ± 6.2 beats/min for partially active recovery after 20-watt pedaling, and 84.4 ± 7.4 beats/min for partially active recovery after 40-watt pedaling.

Figure 1.6b shows the time courses of stroke volume during inactive recovery and two partially active recoveries. Stroke volume during inactive recovery gradually decreased to a constant level, whose mean value at the tenth minute was significantly ($P < 0.05$) less than the baseline value at rest before exercise. On the other hand, during active recovery with light-intensity exercise at 20 and 40 watts, both

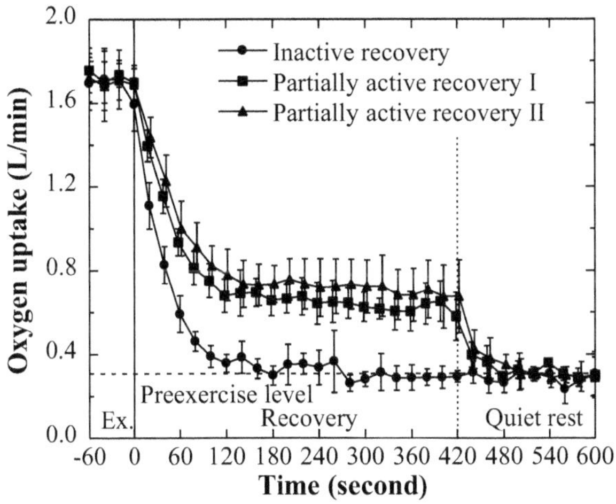

Fig. 1.5 Oxygen uptake during heavy exercise and during inactive recovery and two partially active recoveries. Heavy exercise for 10 min was set at a relative intensity of 68 ± 4% of peak oxygen uptake. Active-recovery exercise for 7 min was set at a relative intensity of 25 ± 2% of peak oxygen uptake and 28 ± 3% of peak oxygen uptake. Data (mean ± SD) are group mean values obtained from five subjects for each of the three experiments. Vertical solid line (at $t = 0$) indicates the start of recovery and vertical dotted line the end of active recovery followed by complete rest in a sitting position on the ergometer. Adapted from Takahashi and Miyamoto (1998), reproduced by permission

stroke volumes were maintained at the levels similar to those during heavy exercise at 170 watts. Both mean stroke volumes at the seventh minute during active recovery were significantly ($P < 0.05$) greater by about 20% than those during inactive recovery. The mean values at quiet rest after the active recovery were also significantly ($P < 0.05$) greater than those during inactive recovery.

Figure 1.6c shows the time courses of cardiac output during inactive recovery and two partially active recoveries. The decreases in cardiac output presented exponential-like manners during inactive and active recovery. The degree of reduction during each active recovery was significantly (each $P < 0.05$) lower than during inactive recovery; in other words, both cardiac outputs remained elevated by about 27% above that of during inactive recovery. Since there was no difference in heart rate, the difference in cardiac output between inactive recovery and each of the two active recoveries was mainly attributed to the difference in stroke volume. During the subsequent rest after active recovery, each cardiac output approached a level similar to that of during inactive recovery.

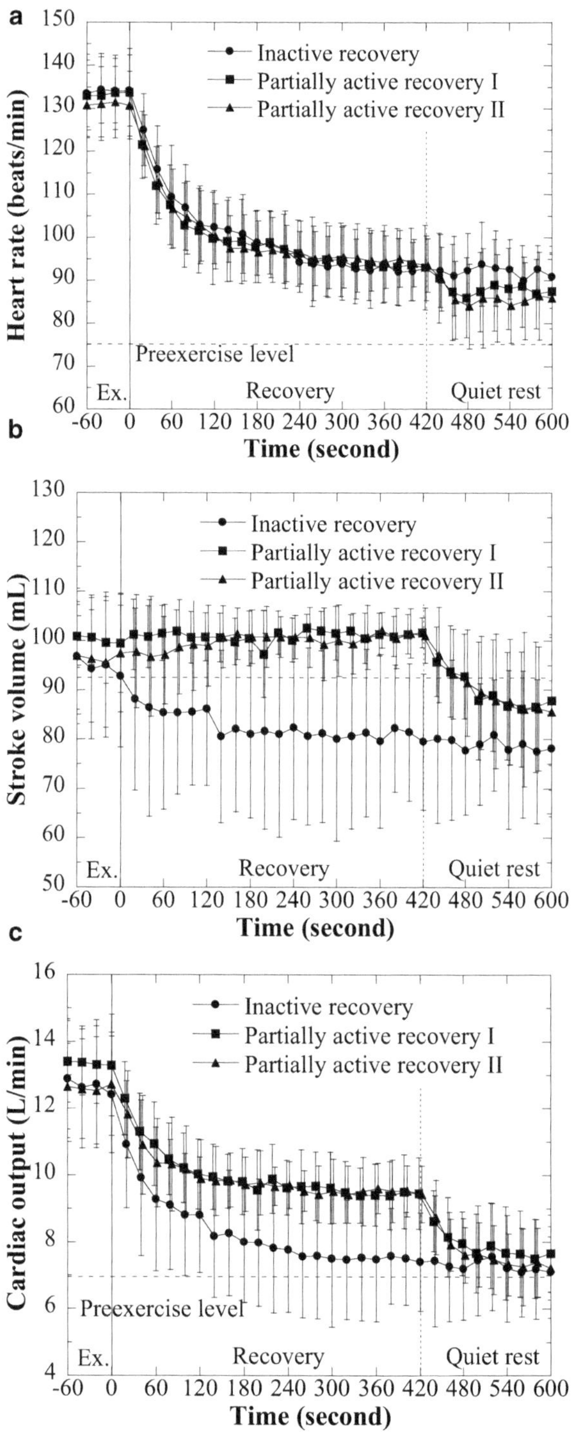

Fig. 1.6 Heart rate (**a**), stroke volume (**b**), and cardiac output (**c**) during heavy exercise and during inactive recovery and two partially active recoveries. Heavy exercise for 10 min was set at a relative intensity of 68 ± 4% of peak oxygen uptake. Active-recovery exercise for 7 min was set at a relative intensity of 25 ± 2% of peak oxygen uptake and 28 ± 3% of peak oxygen uptake. Data (mean ± SD) are group mean values obtained from five subjects for each of the three experiments. Vertical solid line (at $t = 0$) indicates the start of recovery and vertical dotted line the end of active recovery followed by complete rest in a sitting position on the ergometer. Adapted from Takahashi and Miyamoto (1998), reproduced by permission

1.4 Discussion

1.4.1 Differences in Cardiac Recovery

After the same heavy exercise of 10 min at a relative intensity of 68% of peak oxygen uptake, the fast decreases in heart rate during the initial 2-min phase of recovery and their slower decreases during the subsequent phase were observed during inactive recovery and two active recoveries. The time courses of heart rate during the active-recovery period of 7 min were almost the same as those during inactive recovery. However, for 3 min after cessation of active-recovery cycling at relative intensities of 25% and 28% of peak oxygen uptake, the two resting heart rates were found to decrease by 7.5 ± 4.4 and 10.0 ± 3.1 beats/min, respectively; the extent of these reductions was significantly greater than (1.4 ± 2.5 beats/min) during inactive recovery. Stroke volume and cardiac output during the two active recovery modes were significantly higher than those during inactive recovery. By the production of a lower heart rate and a higher stroke volume, the subsequent resting cardiac outputs after both active recoveries were not different from those during inactive recovery.

1.4.2 Stroke Volume and Cardiac Output

The stroke volume was continuously maintained during active recovery after heavy exercise, but that was not the case during inactive recovery. Since there was little difference between the time courses of heart rate reduction for inactive and active recovery, the increased cardiac output during active recovery was mainly attributable to the elevated stroke volume, as compared with inactive recovery. This stroke volume during active recovery was maintained at a level similar to that of during heavy exercise. These results are consistent with the results reported by Goldberg and Shephard, who studied stroke volume during recovery from cycle exercise by using a carbon-dioxide rebreathing method for measurement of cardiac output. On the other hand, a study of other researchers has demonstrated that the decline of stroke volume during inactive recovery to below a pre-exercise resting level is due to the venous pooling of blood in postexercise motionless muscles. Collectively, these results suggest that continuous muscle contractions during active recovery from exercise are important to maintain stroke volume and cardiac output at the resting baseline levels or higher.

1.4.3 Venous Blood Accumulation

The venous pooling in the legs after heavy exercise is due to not only a physical factor (the effect of gravity), but also physiological factors (vasoactive substances produced in the body). The metabolites of exercising muscles, vasodilators, can make the blood vessels dilate at least for the time of their leaving muscle tissues after exercise. There are also neural efferents from the cardiovascular control center to the dilation vessels. However, the blood vessels may be blunt in response to the neural signals of baroreflexes or of chemoreflexes from the active muscles. In such situations, the feedback mechanism of the blood pressure control system may lose fine regulation of the circulation. It is thus stressful for the cardiovascular system that vasodilation is prolonged in the dependent muscles and then arterial blood pressure is of unfavorable instability, thereby leading to postexercise hypotension or syncope after the sudden end of exercise without a cool-down.

1.4.4 Muscle Pump

It has been demonstrated that the muscle blood flow in the lower extremities decreases rapidly after the cessation of calf exercise (rhythmic extension and flexion) in an upright position. Understandably, perfusion pressure in the legs during exercise is increased because venous pressure is decreased by a pumping action of the exercising muscles, called the muscle pump. The perfusion pressure gradient is brought about by rhythmic contraction of the muscles. The blood flow in the active muscles is much maintained, regardless of whether metabolic vasodilation occurs, compared with that in the inactive muscles during recovery from exercise. Our other study has demonstrated that the muscle blood flow after knee extension exercise is significantly greater during active recovery than during passive recovery (involuntary leg movements) and inactive recovery (complete rest). Noticeably, higher stroke volume during active recovery observed in that study can be explained by increased venous return, greater muscle blood flow, and decreased peripheral vascular resistance that are all due to the muscle pump.

1.4.5 Active Recovery Hastens Heart Rate Recovery

Heart rate during inactive recovery from cycle exercise remained elevated above the pre-exercise resting level for at least 10 min, and the result is consistent with the reports of other investigators. However, the mechanisms that underlie this heart rate elevation in late recovery have been poorly understood. Thus far, it has also remained unclear whether recovery light-exercise can enhance the recovery of heart rate from heavy muscular exercise. To our knowledge, our experiment was probably the first to demonstrate that the time course of heart rate during active recovery with light-load cycling is not different from that of during inactive recovery, whereas the subsequent heart rate at complete rest after active recovery promptly decreased to below that of during inactive recovery.

In this study, the recovery period may have been insufficient to observe the slow component of heart rate in the recovery phase, since the heart rates for both partially active recoveries at the tenth minute were still higher than the baseline levels at rest before exercise. To confirm the effective duration of recovery exercise, we examined the heart rate in one additional subject both during 30 min of inactive recovery and during 27 min of active recovery at 40 watts followed by 3 min of rest on the cycle ergometer. As shown in Fig. 1.7, the time course of heart rate during active recovery was almost the same as that of during inactive recovery, despite the difference in beat-to-beat fluctuations. The resting heart rate of post-active recovery after a long-term pedaling could completely return to the baseline level at rest before exercise, whereas the heart rate during inactive recovery remained above the pre-exercise baseline level. These results provide clear evidence that the restoration of heart rate is enhanced effectively by light-intensity exercise during active recovery, as compared with inactive recovery. Therefore, we conclude that light-intensity exercise performed during the recovery period after intensive exercise is useful to shorten the time for the recovery of heart rate to the baseline level at rest before exercise.

1.4.6 Autonomic Nervous System

Several other studies have demonstrated that a prompt increase in heart rate during the transition from rest to light-intensity exercise is mainly because of the withdrawal of efferent vagal tone to the sinoatrial node, in which the intrinsic rhythmicity originates, the so-called cardiac pacemaker. There have also been reviews that neural efferents that separately send the autonomic signals, an increase in sympathetic nerve activity and a decrease in parasympathetic (vagal) nerve activity, to the heart are mediated by a part of the central command that directs the somatic nervous

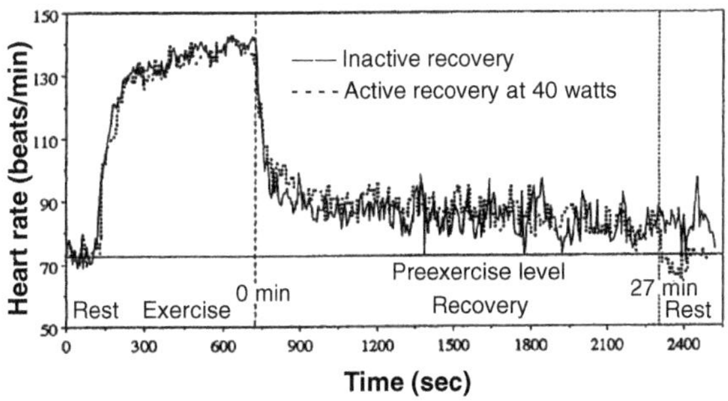

Fig. 1.7 Heart rate (HR) during a 10-min exercise at 170 watts and then during 27 min of inactive and active recovery followed by 3 min of subsequent rest for one subject. Adapted from Takahashi and Miyamoto (1998), reproduced by permission

system to perform exercise. Thus, the control balance of the autonomic nervous system on the cardiac rhythm and contractility during light-intensity exercise is probably different from that at rest; for example, the control of heart rate is dominated by the parasympathetic nerve activity at rest and by the sympathetic nerve activity during exercise.

According to the above, it is likely that light-intensity exercise would more or less effect the elevation of heart rate during active recovery, as compared with inactive recovery. As the exercise during active recovery was performed at a low intensity, less than 33% of peak oxygen uptake, no additional activation of the cardiac sympathetic nervous system would be expected. In this case, heart rate during active recovery would be inevitably increased, above the resting level during inactive recovery, solely due to the withdrawal of vagal nerve activity. The reason is that the parasympathetic nervous system should be depressed in the presence of the central command for performing light exercise. Thus, the difference in the balance of the autonomic nervous system between inactive and active recovery may have caused the separation of the courses of recovery heart rate from each other. However, we found no obvious difference in recovery heart rate between inactive and active recovery. Importantly, the subsequent resting heart rate decreased immediately after the end of light-intensity exercise of active recovery. This reduction in heart rate may be due to an abrupt input of the vagus nerves to the heart. Such vagal activation is likely to require some suppression of vagal activity in the presence of both the central motor command and the neural feedback of the exercising muscle mechanoreflex during active recovery in the background.

1.4.7 Resetting of Baroreceptor Reflexes

For the human cardiovascular control during inactive recovery in a sitting position, the metaboreflex afferent nerves originate from postexercise muscles to the cardiovascular control center, the nucleus of the solitary tract. The metaboreflex changes the operating point of the arterial pressure and heart rate relationship. This neural reflex control of the heart rate is attributed to sympathetic nerve activation elicited mainly by the group IV afferents of metaboreceptors (chemoreceptors) in the lower-limb muscles, in which an amount of venous blood would be accumulated in the absence of the muscle pump. Some studies in dogs and in humans have shown that during inactive recovery from exercise, the changes in heart rate with postexercise muscle ischemia are similar to those without the ischemia. In contrast, systemic arterial pressure is different between with and without muscular perfusion; that is, arterial blood pressure remains elevated in response to the ischemia, as compared with the control perfusion. Another study has also examined autonomic control of the postexercise human hearts that were denervated pharmacologically by blockades of vagal control with atropine (muscarinic receptor antagonist) and of sympathetic control with adrenergic ß-receptor antagonist. The results have demonstrated that the postexercise reduction of heart rate is mediated primarily by increased vagal nerve activity to the heart, despite the fact that the sympathetic nerve activity is reflexly sustained via the neural afferents originating from the muscular metaboreceptors.

Alternatively, it can be argued that the shift of the reference of reflex heart rate to lower arterial pressure during recovery from exercise would be mediated by baroreceptor resetting in the cardiovascular regulation, whose physiological function may be related to postexercise hypotension. Unfortunately, the current study did not measure any activity of the two divisions of the autonomic nervous system, blood catecholamine concentrations, or blood pressure. Thus, further study is needed to elucidate the mechanism involved in the elevation of recovery heart rate for a relatively long period. These relevant issues to baroreflex function will be discussed in more detail in Chap. 5.

1.4.8 Antiarrhythmic Effect of Active Recovery

It has been reported that during inactive recovery from exercise, the return of plasma potassium ion (K^+) to the prevailing concentration level before exercise is faster than that of plasma norepinephrine. It has also been noted that hyperkalemia can abolish or reduce the incidence of norepinephrine-induced arrhythmias. After exercise, the plasma concentration of norepinephrine decreased more slowly from the level at exercise, because the sympathetic nerve activity is not simply decreased or is sustained with the neural metaboreflex of chemoreceptors within the postexercise muscles. The concentration unbalanced conditions may thus be a risk factor for the development of cardiac arrhythmias during the postexercise period, especially inactive recovery. Importantly, the relevant literature has suggested that vagal rebound in the recovery phase from exercise probably minimizes the potentially proarrhythmic effects of hypokalemia and high sympathetic tone. This suggestion is supported partly by the findings that decreased sympathetic tone, increased parasympathetic tone, or the predominance of the cardiac vagal tone over the sympathetic tone in active individuals with increased levels of physical activity, exercise training, and fitness would result in significant protection from sudden cardiac death. Accordingly, we speculate that the physiological effects of postexercise recovery exercise at light or mild intensity on cardiac recovery can involve several factors as follows: (1) the antiarrhythmic effect of mutual antagonism between plasma potassium ion and norepinephrine levels on the heart may be maintained during the early phase of recovery, because the reduction of blood potassium ions should be delayed by continuous muscle contraction during active recovery; and (2) light- or mild-intensity exercise may have a somewhat restorative effect on the autonomic nervous system with an increase in cardiac vagal tone, a decrease in cardiac sympathetic tone, or both, which the autonomic effects would reflect in the slight, but significant decrease in the heart rate after the end of active-recovery exercise in this study (also see Chap. 3).

These physiological effects of active-recovery exercise on the recovery heart rate, stroke volume, and cardiac output from heavy exercise have not been extensively explored so far. Nevertheless, a study has at least suggested that the best strategy is for the exercise intensity to be diminished gradually, but the worst strategy for recovery from exercise would be possibly to stop exercising abruptly. It has also been suggested that an extended cool-down exercise, as well as a gradual warm-up, probably plays a role in protection against cardiac catastrophe, especially in hot environments.

1.5 Summary

The cardiac responses to light-intensity exercise during active recovery from a heavy cycle exercise were examined in five male college students in an upright seated position. A significant effect of light exercise was apparent specifically on the facilitation of heart rate recovery to the pre-exercise resting level during the rest period after active recovery. The active recovery exercise also played an important role in preventing the venous pooling in the metabolically dilated vasculature of the postexercise muscles. Then, muscular contraction, the muscle pump, during active recovery can facilitate the venous return of blood to the heart. Central blood volume increased by active-recovery exercise in the upright position would contribute to the maintenance of both stroke volume and cardiac output above the baseline levels at rest before exercise. However, it remains to be elucidated whether the interplay between the parasympathetic and sympathetic nervous systems would be implicated in the mechanism for the relatively stable heart rate elevated above the resting baseline level during recovery. It must also be noted that such factors as muscular metaboreflex, arterial baroreflex control, and changes in the concentrations of plasma catecholamines should contribute to the modulation of the cardiovascular recovery responses. Further studies are necessary to determine the effects of the magnitude and duration of recovery exercise on reductions in heart rate during active recovery. Further research will provide important information about the protocols for recovery of heart rate and for prevention of sudden cardiac death after exercise.

Further Reading

Takahashi T, Miyamoto Y. Influence of light physical activity on cardiac responses during recovery from exercise in humans. Eur J Appl Physiol. 1998;77:305–11.

Bangsbo J, Graham T, Johansen L, Saltin B. Muscle lactate metabolism in recovery from intense exhaustive exercise: impact of light exercise. J Appl Physiol. 1994;77:1890–5.

Baum K, Eβfeld D, Leyk D, Stegemann J. Blood pressure and heart rate during rest-exercise and exercise-rest transitions. Eur J Appl Physiol. 1992;64:134–8.

Billman GE, Schwartz PJ, Stone HL. The effects of daily exercise on susceptibility to sudden cardiac death. Circulation. 1984;69:1182–9.

Boer P, Roos JC, Geyskes GG, Dorhout Mees EJ. Measurement of cardiac output by impedance cardiography under various conditions. Am J Phys. 1979;237:H491–6.
Brenner IKM, Thomas S, Shephard RJ. Spectral analysis of heart rate variability during heat exposure and repeated exercise. Eur J Appl Physiol. 1997;76:145–56.
Brooks GA, Fahey TD, Baldwin KM. Exercise physiology: human bioenergetics and its application. New York: McGraw-Hill Companies, Inc.; 2005.
Cybulski G. Ambulatory impedance cardiography: the systems and applications. Berlin: Springer-Verlag; 2011.
Dimsdale JE, Hartley LH, Guiney T, Ruskin JN, Greenblatt D. Postexercise peril: plasma catecholamines and exercise. JAMA J Am Med Assoc. 1984;251:630–2.
Dodd S, Powers SK, Callender T, Brooks E. Blood lactate disappearance at various intensities of recovery exercise. J Appl Physiol. 1984;57:1462–5.
Floras JS, Sinkey CA, Aylward PE, Seals DR, Thoren PN, Mark AL. Postexercise hypotension and sympathoinhibition in borderline hypertensive men. Hypertension. 1989;14:28–35.
Fowler KT. The respiratory mass spectrometer. Phys Med Biol. 1969;14:185–99.
Goldberg DI, Shephard RJ. Stroke volume during recovery from upright bicycle exercise. J Appl Physiol. 1980;48:833–7.
Guyton AC, Douglas BH, Langston JB, Richardson TQ. Instantaneous increase in mean circulatory pressure and cardiac output at onset of muscular activity. Circ Res. 1962;11:431–41.
Hausswirth C, Mujika I. Recovery for performance in sport. Champaign, Illinois: Human Kinetics; 2013.
Hughson RL, Xing HC, Borkhoff C, Butler GC. Kinetics of ventilation and gas exchange during supine and upright cycle exercise. Eur J Appl Physiol. 1991;63:300–7.
Krogh A. The comparative physiology of respiratory mechanisms. Philadelphia: University of Pennsylvania Press; 1941.
Kubicek WG, Karnegis JN, Patterson RP, Witsoe DA, Mattson RH. Development and evaluation of an impedance cardiac output system. Aerosp Med. 1966;37:1208–12.
Kubicek WG, Patterson RP, Witsoe DA. Impedance cardiography as a noninvasive method of monitoring cardiac function and other parameters of the cardiovascular system. Ann N Y Acad Sci. 1970;170:724–32.
Laughlin MH. Skeletal muscle blood flow capacity: role of muscle pump in exercise hyperemia. Am J Phys. 1987;253:H993–H1004.
Van Leeuwen BE, Barendsen GJ, Lubbers J, De Pater L. Calf blood flow and posture: Doppler ultrasound measurements during and after exercise. J Appl Physiol. 1992;72:1675–80.
Linarsson D. Dynamics of pulmonary gas exchange and heart rate changes at start and end of exercise. Acta Physiol Scand Suppl. 1974;415:1–68.
Maehlum S, Grandmontagne M, Newsholme EA, Sejersted OM. Magnitude and duration of excess postexercise oxygen consumption in healthy young subjects. Metabolism. 1986;35:425–9.
Menzies P, Menzies C, Mcintyre L, Paterson P, Wilson J, Kemi OJ. Blood lactate clearance during active recovery after an intense running bout depends on the intensity of the active recovery. J Sports Sci. 2010;28:975–82.
Miyamoto Y, Takahashi M, Tamura T, Nakamura T, Hiura T, Mikami T. Continuous determination of cardiac output during exercise by use of impedance plethysmography. Med Biol Eng Comput. 1981a;19:638–44.
Miyamoto Y, Tamura T, Mikami T. Automatic determination of cardiac output using an impedance plethysmography. Biotelem Patient Monit. 1981b;8:189–203.
Mohapatra SN. Non-invasive cardiovascular monitoring by electrical impedance technique. London: Pitman Medical; 1981.
Niizek K, Kawahara K, Miyamoto Y. Interaction among cardiac, respiratory, and locomotor rhythms during cardiolocomotor synchronization. J Appl Physiol. 1993;75:1815–21.
O'Leary DS. Autonomic mechanisms of muscle metaboreflex control of heart rate. J Appl Physiol. 1993;74:1748–54.

Orizio C, Perini R, Comande A, Castellano M, Beschi M, Veicsteinas A. Plasma catecholamines and heart rate at the beginning of muscular exercise in man. Eur J Appl Physiol. 1988;57:644–51.
Paterson DJ. Antiarrhythmic mechanisms during exercise. J Appl Physiol. 1996;80:1853–62.
Perini R, Orizio C, Comnde A, Castellano M, Beschi M, Veicsteinas A. Plasma norepinephrine and heart rate dynamics during recovery from submaximal exercise in man. Eur J Appl Physiol. 1989;58:879–83.
Rowell LB, Brengelmann L, Blackmon JR, Bruce RA, Murray JA. Disparities between aortic and peripheral pulse pressures induced by upright exercise and vasomotor changes in man. Circulation. 1968;37:954–64.
Rowell LB, O'Leary DS. Reflex control of the circulation during exercise: chemoreflexes and mechanoreflexes. J Appl Physiol. 1990;69:407–18.
Siebenmann C, Rasmussen P, Sørensen H, Zaar M, Hvidtfeldt M, Pichon A, Secher NH, Lundby C. Cardiac output during exercise: a comparison of four methods. Scand J Med Sci Sports. 2015;25:e20–7.
Stick C, Jaeger H, Witzleb E. Measurements of volume changes and venous pressure in the human lower leg during walking and running. J Appl Physiol. 1992;72:2063–8.
Takahashi T. Easy to understand data analysis and statistics: let's start analysis statistics of medical and life science with excel. Tokyo: Ohmsha, Ltd.; 2017.
Takahashi T, Niizeki K, Miyamoto Y. Effects of base line changes in work rate on cardiorespiratory dynamics in incremental and decremental ramp exercise. Adv Exp Med Biol. 1995;393:159–64.
Takahashi T, Niizeki K, Miyamoto Y. Respiratory responses to passive and active recovery from exercise. Jpn J Physiol. 1997;47:59–65.
Takahashi T, Okada A, Tamura T, Miyamoto Y. Cardiorespiratory responses to walking and running at an incremental speed of treadmill. Adv Exp Med Biol. 2001;499:363–8.
Takahashi T, Saitoh T, Okada A, Matsuo T. Differences in femoral artery blood velocity among active, inactive, and passive recovery modes following knee extension and flexion exercise. Ther Res. 2006;27:1393–403.
Takahashi T, Yanagida H, Niizeki K, Wang S, Matsuo T, Watanabe S, Okada A, Kuwayama T, Natsui J. Economical estimates of oxygen uptake as a function of gait parameters for an ambulatory monitoring system. Ther Res. 2009a;30:219–25.
Takahashi T, Yanagida H, Wang S, Matsuo T, Okada A, Kuwayama T, Natsui J. Minimum oxygen cost of human walking with geometrically similar leg movements. Ther Res. 2009b;30:113–24.
Victor RG, Seals DR, Mark AL. Differential control of heart rate and sympathetic nerve activity during dynamic exercise. Clin Invest. 1987;79:508–16.
Williams ME, Gervuno EV, Rosa RM, Landsberg L, Young JB, Silva P, Epstein FH. Catecholamine modulation of rapid potassium shifts during exercise. N Engl J Med. 1985;312:823–7.
Yoshiya I, Nakajima T, Nagai I, Jitsukawa S. A bidirectional respiratory flowmeter using the hot-wire principle. J Appl Physiol. 1975;38:360–5.
Waaler BA, Eriksen M, Toska K. The effect of meal size on postprandial increase in cardiac output. Acta Physiol Scand. 1991;142:33–9.

Chapter 2
Cardiovascular Recovery in Different Postures

2.1 Introduction

After exercise, heart rate in the recovery phase shows an initial rapid decrease and then a gradual decrease to a relatively stable level. The recovery heart rate remains elevated above the pre-exercise resting level for longer than expected. In general, as the body position of resting humans changes from an upright to a supine position, the sympathetic nerve activity outflow to the heart decreases and the vagal nerve activity outflow increases. Similarly, if heart rate is under the autonomic control during recovery from exercise, elevated heart rate during recovery would be reduced with the predominance of vagus tone in supine posture compared with upright posture.

On transition from a supine to an upright position, the distribution of blood volume in the body is affected profoundly by gravity. The control mechanisms of baroreflexes are involved in the cardiovascular system to compensate for the deviation of mean blood pressure values from a set point during posture changes. Numerous studies have demonstrated that after dynamic exercise, arterial blood pressure is substantially decreased below its pre-exercise resting level, the so-called postexercise hypotension, which occurs regardless of body position. However, the direct and indirect neurogenic pathways by which the decreases in arterial blood pressure as the aftereffect of exercise may be regulated are incompletely understood. Our poor understanding of detailed mechanisms of postexercise hypotension would be attributable to relative few studies of the differences in blood distribution in the body between lying and sitting positions during recovery from exercise.

Until the present, except some studies, little research has focused on the intriguing questions as to how changes of body position affect the dynamics of cardiovascular responses during recovery from exercise. A study with a Doppler ultrasound technique has demonstrated that stroke volume is significantly increased during recovery in a supine position, whereas it is decreased during recovery in an upright

T. Takahashi, *The Cardiovascular System During Exercise and Recovery*,
https://doi.org/10.1007/978-981-99-8486-2_2

position, both after cycle ergometer exercise in the same upright position. In that study, a general description offered us to interpret the hemodynamic responses related to changes in stroke volume. However, it has not dealt adequately with the integrated control mechanisms of the circulatory system that underlie the complicated cardiovascular adjustments to postexercise recovery dependent on the different postures.

The objective of this chapter is, therefore, to address the open question of the cardiovascular control mechanisms for postexercise arterial blood pressure in different body positions. For attaining this objective, we will examine the changes in cardiovascular variables during recovery in upright and supine positions after exercise performed in the same upright position. During the period of recovery, subjects rest in either an upright sitting or a supine lying position after they perform the upright cycle exercise at a heavy work load. In the experiments, we continuously measured heart rate, stroke volume, cardiac output, arterial blood pressure, total peripheral vascular resistance, and the amplitude of a high-frequency component of R-R interval variability. Spectrum analysis of R-R interval variability is used to assess the contribution of the cardiac parasympathetic nervous system (vagal nerve efferent activity) to the heart at rest both before and after exercise.

2.2 Materials and Methods

2.2.1 *Human Subjects*

Seven healthy male subjects participated in this study. The physical characteristics of them with a mean age of 22.4 ± 2.1 years (mean ± SD) were as follows: a mean height of 1.70 ± 0.06 m, a mean body mass of 61.1 ± 9.1 kg, and a mean peak oxygen uptake of 2.726 ± 0.400 L/min. All the participants were fully informed of the procedures, risks, and benefits of the study after a medical examination. They gave written informed consent before the study. None of them had any history of circulatory or respiratory disease. They had not been engaged in either regular physical activity or competitive sports. The experimental protocol was approved by the Health Research Ethics Board at the university. This study was carried out in accordance with the principles outlined in the Declaration of Helsinki.

2.2.2 *Respiratory Measurements*

Oxygen uptake (in liters per minute) was measured breath-by-breath with a computerized measurement system. Subjects wore a face mask, the front entrance of which had a hot-wire transducer. Inspiratory and expiratory gases can flow bidirectionally through the hot-wire transducer. The electrical signals produced by the transducer in terms of breathing flow were transmitted to a pneumotachograph (RF-2; Minato, Osaka, Japan); that is, air taken into the lungs and out again from the mouth were monitored as ventilatory airflow with this pneumotachograph. Small samples of the gases from an outlet pipe of side wall of the face mask were introduced into a medical mass spectrometer (WSMR-1400; Westron, Chiba, Japan) via a narrow-bore tube. The composition of breaths was continuously analyzed with the mass spectrometer. The spectrometer was calibrated against a standard reference gas mixture before each measurement.

2.2.3 *Cardiac Measurements*

Heart rate (*HR*, in beats per minute), stroke volume (*SV*, in milliliters per beat), and cardiac output (*CO*, in liters per minute) were measured with a computerized measurement system. R-R (interbeat) intervals were measured with the use of an electrocardiograph (ECG) with three bipolar chest leads. Heart rate was calculated as the reciprocal of R-R intervals multiplied by 60. An impedance cardiograph (ICG) with four electrodes (RGA-5; Nihon Koden, Tokyo, Japan) was used to monitor transthoracic impedance bioelectrically for determining stroke volume noninvasively. For measurement of the bioimpedance, the two circumferential strip electrodes (leads 1 and 2), 5-mm width and 0.5-mm thickness, of copper were placed 30 mm apart at the neck and the other two (leads 3 and 4) were placed at the abdomen, as shown in Fig. 2.1. It is best to place the electrode 2 around the base of the neck and the electrode 3 around the thorax at a level below the xiphoid process. A constant, sinusoidal alternating current of 350 microamperes at a frequency of 100 kilohertz was fed into the two outer electrodes 1 and 4. The electrical impedance between the two inner electrodes 2 and 3 was continuously recorded.

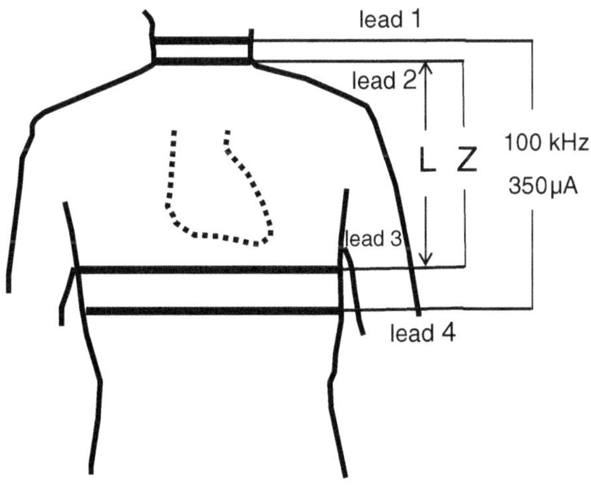

Fig. 2.1 Typical positions of four circumferential impedance electrodes. A constant sinusoidal alternating current is introduced through leads 1 and 4. Leads 2 and 3 detect changes in voltage. *L* stands for the distance between inner-receiving electrodes. ΔZ is the variation of impedance signal

Stroke volume was calculated from the standard equation of Kubicek:

$$SV = \rho\left(L / Z_0\right)^2 \left(dZ / dt\right)_{\min} T_{\mathrm{ET}} \tag{2.1}$$

where ρ (ohm·cm) was the resistivity of blood, L (cm) was the arithmetic mean of the anterior and posterior distances between the inner electrodes, Z_0 (ohm) was the baseline impedance between the two inner electrodes, $(dZ/dt)_{min}$ (ohm/s) was the minimum of the first derivative of the change in thoracic impedance (ΔZ) with respect to time, and T_{ET} (s) was the ventricular ejection time obtained from the *dZ/dt* waveform. The relationships among these parameters in impedance waveforms are shown in Fig. 2.2.

T_{ET} was measured in time from the last zero-line crossing of *dZ/dt* before the $(dZ/dt)_{min}$ to the first positive peak following the $(dZ/dt)_{min}$. Theoretically, the starting point for the determination of T_{ET} was a representative better than the use of 15% of the $(dZ/dt)_{min}$; however, Kubicek et al. have proposed that $0.15{\cdot}(\mathrm{dZ/dt})_{min}$ was still a reliable point for determining T_{ET}, because it could avoid a difficulty in making the start point of T_{ET} from a small oscillation of the waves.

The electric resistivity of blood was estimated with hematocrit (*Hct*, in percent) from a formula:

$$\rho = 66\left(3 + 0.019Hct\right) / \left(3 - 0.038Hct\right) \tag{2.2}$$

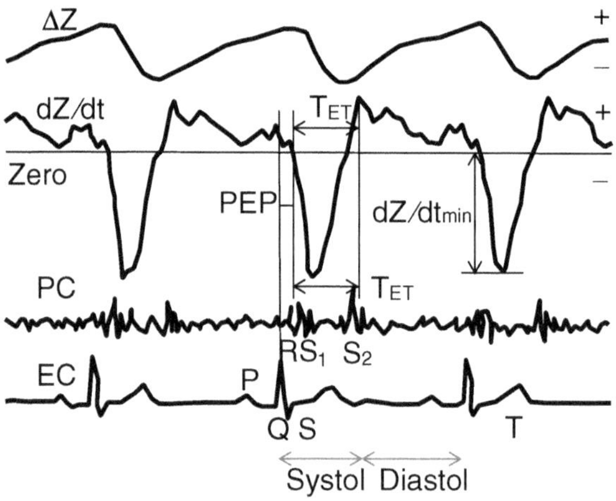

Fig. 2.2 A set of recordings in a normal subject by impedance cardiography: *ΔZ* the variation of impedance signal, *dZ/dt* the first derivative of ΔZ, *the* dZ/dt_{min} the minimum *dZ/dt*, *PEP* pre-ejection period, T_{ET} left ventricular ejection time, *PCG* phonocardiogram, *ECG* electrocardiogram. A decrease in impedance is recorded as a downward deflection in ΔZ and *dZ/dt* tracings. Note that the QRS complex (depolarization of the ventricles) occurs at the beginning of systole and the rise in ventricular pressure. The T wave (repolarization of the ventricles) occurs at the same time that the ventricles relax at the beginning of diastole

where *Hct* was the percent of volume of red blood cells occupying a blood sample.

The ECG and ICG electrodes attached to the surface of the body did not affect spontaneous breathing at rest and during exercise. Some noises of impedance signals caused by both breathing and body motion were eliminated by two mathematical techniques: (1) moving average was the average of ICG waveforms over 10 cardiac cycles, and (2) adding average was the average of two measurements obtained from two trials of the same exercise and recovery tests in each individual. Although absolute values for stroke volumestroke volume derived from impedance signal remain slightly unsettled, there has been a general agreement that ICG measurement is reliable for estimating relative changes in stroke volume from rest to exercise and recovery. Furthermore, the validity of stroke volume estimations with this system has been confirmed under a variety of experimental conditions including rest, exercise, and postural changes. As shown in Fig. 2.1 and Eq. 2.1, the stroke volume measured is a function of the distance between the two inside electrodes. We naturally localized ICG electrodes at the same positions for each individual in order to minimize variations in stroke volume measurements from day to day.

2.2.4 Blood Pressure Measurements

Systolic arterial blood pressure (*SBP*, in millimeters of mercury, mmHg), mean blood pressure (*MBP*), and diastolic blood pressure (*DBP*) were noninvasively measured beat-by-beat with the use of a photoplethysmograph (2300 Finapres BP monitor; Omeda, Tokyo, Japan). This photoplethysmograph has several features: Firstly, a small inflatable finger cuff is connected to a small front-end box attached to the left wrist. Secondly, the finger cuff is located at the middle phalanx of the third finger of the left hand. Thirdly, the change in the blood volume under the cuff is detected by infra-red sensors. Finally, a servo system of the cuff with air supply pump counteracts the intravascular blood volume immediately by increasing or decreasing cuff pressure to the same extent as finger intra-arterial pressure changes. The device thus enables us to monitor beat-to-beat changes in cuff pressure that parallel those in finger intra-arterial blood pressure. Some studies have reported that simultaneous direct and indirect measurement values are similar, but not identical. The differences between the finger cuff and intra-arterial pressures for a group of subjects usually average less than 5 mmHg. Physiologically, the values for finger intra-arterial pressure slightly differ from the values obtained more proximally, since the outline of the pulse wave becomes sharper as it goes to the periphery. In addition, the values for systolic blood pressure measured with the finger cuff are a little higher than those for finger intra-arterial pressure. In contrast, diastolic pressure values of the finger cuff are somewhat lower than those of intra-artery. As a result, the measurement of mean blood pressure by this method will usually be slightly lower than the true values in the intra-artery at the same measuring site. Therefore, it is necessary to interpret pressure values in the finger artery, which are slightly lower than those in the brachial artery.

Mean blood pressure was calculated as diastolic pressure plus one-third of pulse pressure: $MBP = (SBP - DBP)/3 + DBP$. Finapres recordings have been validated by a detailed comparison with the recordings of intra-arterial blood pressure under various conditions. The finger cuffed around the middle phalanx position was kept at heart level to avoid the gravitational influence of blood, termed hydrostatic pressure, from intravascular pressure throughout all the experiments.

Total peripheral vascular resistance (*TPR*, in dyn·sec^{-1} cm^{-5}) was calculated by dividing mean blood pressure by cardiac output and multiplying by 80, where 1 m mHg L^{-1}·min = 80 dyn s^{-1} cm^{-5}. The constant 80 is a conversion factor to change units to standard resistance units.

2.2.5 Power Spectral Analysis

The ECG signal was digitized at a sampling rate of 500 hertz (Hz) according to the literature, which reviews that a series of QRS-complex can be recognized precisely by sampling at an optimal rate of 250–500 Hz. The temporal positions of all R-wave peaks were identified by a peak detection algorithm. An error in R-wave recognition, such as an ectopic beat, if any, was checked and edited manually. No subjects presented any ectopic beats in the experiments. Temporal locations of R-wave peaks determined at uneven time intervals were regularly interpolated with the use of a cubic Lagrange interpolation formula into a series of beat-to-beat R-R intervals. The result of the interpolation was then resampled at 1 Hz to obtain an equidistantly sampled R-R interval function.

The two main spectral components of heart rate variability were distinguished in a spectrum calculated from a short-term recording of 2 min. Spectral analysis was performed on the R-R interval function during the last 2-min period of rest both before and after exercise in the supine and upright positions. There were no marked differences in the average of R-R intervals between the first 30 and last 30 s of the 2-min period in either body position; specifically, the differences were less than 5% for all the recorded data. However, a linear trend, if any, was removed from the R-R interval time series before spectral analysis. This mathematical procedure was to remove very low-frequency power components associated with a long-term biological variation, for example, the circadian rhythm.

The power spectral density derived from the time series of fluctuations in R-R intervals was assessed with a maximal entropy method operated by a software installed in a commercially available computerized system (MemCal, Suwa-Torasuto, Tokyo, Japan). The validity of this spectral analysis system has been confirmed in a human study with regard to R-R interval data.

The magnitude of a high-frequency component (*HF*, 0.15–0.40 Hz) of R-R interval variability was assessed as a change in efferent vagal nerve activity to the heart. A low-frequency component (LF, 0.04–0.15 Hz) was not analyzed because there is still some ambiguity about the physiological background of the low-frequency component. The magnitude of the spectral high-frequency component was expressed as mean amplitude ($[2 \times \text{power of component}]^{1/2}$). The data of R-R intervals obtained during exercise were excluded from spectral analysis, since an unsolved problem remains in the method of spectral analysis to evaluate the autonomic traffic to the heart during exercise. The simple reason why it is difficult to analyze R-R intervals is that fluctuations of heart rate weaken during intensive exercise.

2.2.6 Preliminary Test

A week before the first exercise test, each subject in a seated upright position performed a ramp-exercise test with an incremental work rate of 15 watts every minute for determination of peak oxygen uptake. The exercise was conducted on a stationary cycle ergometer with an electromagnetic brake resistance (Corival 300; Lode, The Netherlands) until the subject could not continue to cycle exercise any more. The end point termed "exhaustion" against exercise was taken as a point when the subject was unable to maintain a pedaling rate of at least 50 revolutions per minute. The peak oxygen uptake was defined as the highest oxygen uptake obtained as the mean value over a 10-s period of breath-by-breath measurements; in general, the term peak oxygen uptake is used when the test performance appears to be limited by local muscular factors rather than the function of the central circulatory system.

2.2.7 Experimental Protocols

The subjects underwent two types of recovery posture after they performed the same heavy exercise in an upright sitting position (Fig. 2.3). Each of the two postures included either an upright (sitting) position or a supine (lying) position in which they rested during a recovery period after heavy exercise.

Before performing the exercise, the subjects reclined on a bed for 15 min and further rested in an upright seated position on the cycle ergometer for 5 min. They performed a warm-up exercise of unloaded pedaling for 3 min in the upright position and then a stimulus exercise at a constant intensity corresponding to 80% of peak oxygen uptake for 5 min in the same position. After the exercise, they completely rested in a motionless state during a 10-min period of recovery in each of the two different positions: the sitting position on the ergometer or the lying position on the bed next to the ergometer. In the supine recovery, the subjects completed the postural change from upright to supine within 15 s after exercise. Since multiple electrodes had been attached to their body, such time was needed to prevent the sensors from falling out.

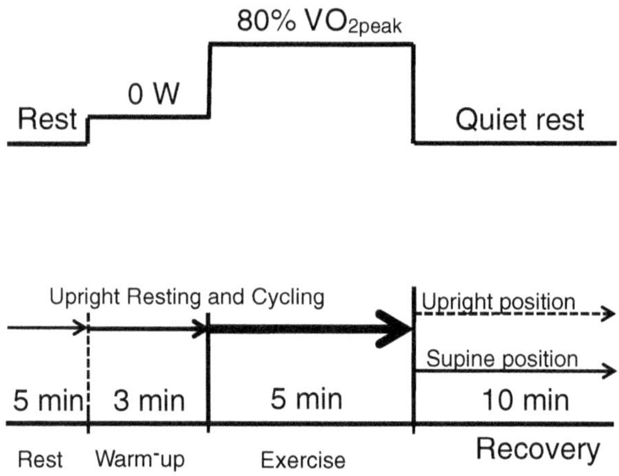

Fig. 2.3 Schematic diagram of exercise protocols. The subjects rested in a lying position for 15 min and further in a seated position on the ergometer for 5 min. They performed a warm-up exercise at a work rate of 0 watts for 3 min and then stimulus exercise at a work rate of 80% of peak oxygen uptake for 5 min in the upright position. After exercise, they underwent quietly resting recovery during a 10-min period in either the upright or the supine position

During exercise, the subjects pedaled at a constant rate of 60 revolutions per minute and maintained the pace in synchrony with an audible metronome. They placed their feet on the footrest near the flywheel during the rest periods in the upright position. No attempt was made to regulate the breathing pattern, depth, or frequency throughout the experiments; in short, each person breathed naturally so that the respiratory and circulatory systems would not be subjected to nonphysiological conditions.

2.2.8 Experimental Schedule

All the exercise tests were conducted between 10:00 and 17:00. The time of days at which the tests were performed varied from subject to subject; however, each subject underwent them at almost the same time of the days. The subjects were scheduled to participate in the experiments 1 day a week; the interval of which would leave no effect of exercise training on the cardiorespiratory responses to exercise and recovery. They were instructed to eat a light meal at least 3 h before the tests, but not to consume caffeine or alcohol at least 24 h before the tests. All experiments were conducted at a room temperature of 19–23 °C (degrees Celsius).

2.2.9 Data Analysis

Cardiorespiratory data obtained from two repetitions of each test for each subject were regularly interpolated at a time resolution of five-second intervals by using the Lagrange interpolation. The two interpolated data were then averaged to yield a single set of data for each subject. Group mean values were obtained by averaging the sets of data for seven individuals. Mean values for all variables (except high-frequency amplitude) were determined from data for the last minute before, during, and after exercise. The amplitude of the high-frequency component of R-R interval variability was determined from data recorded for the last 2 min before and after exercise.

2.2.10 Statistical Analysis

Two-way factorial analysis of variance was used to assess differences in the effects of upright and supine positions on mean values for all variables before, during, and after exercise. As a post hoc test, two-sided Student's t test for paired samples was used to determine significant differences. A least-squares regression line was fitted to the mean R-R interval against the high-frequency amplitude with the individual average data together before and after exercise in both positions. A correlation between the mean R-R interval and the high-frequency amplitude was statistically examined with the use of the Pearson product-moment correlation coefficient. A P value of less than 0.05 was considered significant in all statistical analyses. Data are presented as mean ± standard deviation (SD).

2.3 Results

When the subjects changed the body from a supine to an upright sitting position at rest before exercise, stroke volume and cardiac output decreased significantly by about 40% ($P < 0.05$) and 20% ($P < 0.05$), respectively. At the time, heart rate increased significantly by about 30% ($P < 0.05$) on average. The decrease in cardiac output was accompanied by an increase in total peripheral resistance in the upright position, thereby leading to no significant difference in resting mean blood pressure from the supine position. Table 2.1 summarizes the average values of all the test variables obtained for seven subjects, who underwent the two types of recovery (upright recovery vs. supine recovery) after the same rest and exercise in the upright position. The constant intensity of upright cycle exercise for 5 min was equivalent to 80% of peak oxygen uptake. The mean values for each variable during both rest and exercise periods were not statistically different between the two experiments of subsequent recovery different positions.

Table 2.1 Cardiorespiratory variables before, during, and after exercise

	f (breaths/min)	*HR* (beats/min)	*SV* (mL)	*CO* (L/min)	*SBP* (mmHg)	*MBP* (mmHg)	*DBP* (mmHg)	*TPR* (dyn·s^{-1}·cm^{-5})	*HF* (ms)
Upright recovery experiment									
Upright rest	14.5 ± 5.6	81.7 ± 8.6	64.9 ± 15.5	5.3 ± 0.8	126.1 ± 10.0	89.7 ± 9.6	71.5 ± 10.0	1433.3 ± 418.7	30.5 ± 14.7
Upright exercise	28.4 ± 4.7	150.9 ± 12.1	71.3 ± 20.3	10.6 ± 2.4	198.1 ± 15.4	126.5 ± 10.5	90.7 ± 8.6	1012.2 ± 294.9	–
Upright recovery	17.6 ± 3.9	95.0 ± 11.0*	57.0 ± 13.8	5.3 ± 0.9	123.8 ± 12.5	90.1 ± 9.6	73.2 ± 9.1	1418.5 ± 416.1	15.1 ± 10.0*
Supine recovery experiment									
Supine rest	12.4 ± 4.6	62.3 ± 6.9†	107.7 ± 14.9†	6.7 ± 0.7†	123.6 ± 14.1	87.1 ± 10.5	68.9 ± 11.9	1032.4 ± 145.8†	58.6 ± 15.9†
Upright exercise	28.2 ± 5.0	151.2 ± 11.6	75.2 ± 15.8	11.2 ± 1.6	199.1 ± 17.0	127.1 ± 7.6	91.0 ± 5.3	929.1 ± 169.6	–
Supine recovery	16.3 ± 2.5	81.4 ± 6.7*†	101.2 ± 20.8†	8.2 ± 1.5*†	128.0 ± 14.1	88.9 ± 10.1	69.3 ± 9.2	907.6 ± 245.2*†	22.8 ± 11.4*

Values are expressed as mean ± SD for seven subjects. Values for supine rest, upright rest, exercise, and recovery were obtained at the fifth minute of supine rest, the fifth minute of upright rest, the fifth minute of exercise, and the tenth minute of recovery, respectively

f respiratory frequency, *HR* heart rate, *SV* stroke volume, *CO* cardiac output, *TPR* total peripheral vascular resistance, *HF* amplitude of high-frequency component of R-R interval variability

$^{*}P < 0.05$, significantly different from the resting value for the same posture. $^{\dagger}P < 0.05$, significantly different from the corresponding value for upright posture. Adapted from Takahashi et al. (2000), reproduced by permission

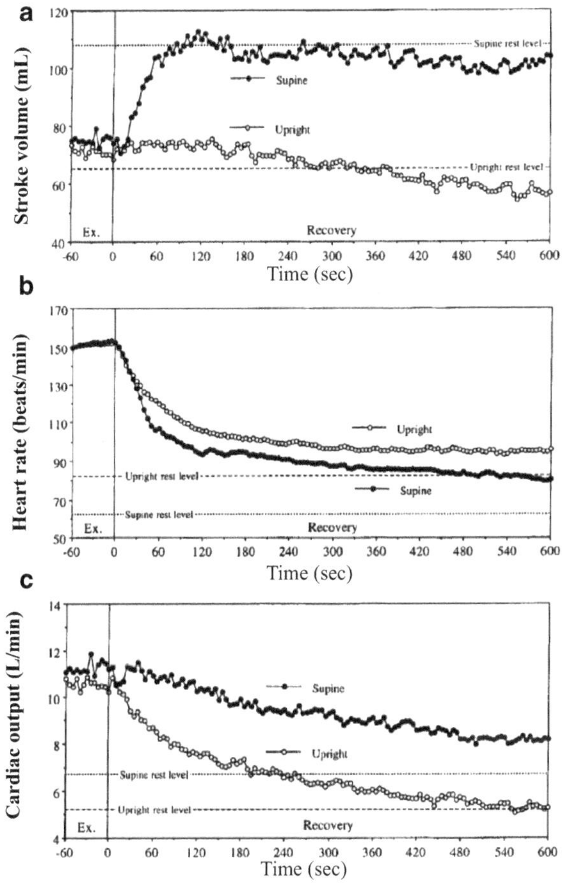

Fig. 2.4 Stroke volume (**a**), heart rate (**b**), and cardiac output (**c**), mean blood pressure (**d**), and total peripheral resistance (TPR, **e**) during a 10-min period of recovery from upright exercise at relative intensity of 80% of peak oxygen uptake. Data (open circles for upright position and close circles for supine position) are group mean values obtained from seven male subjects for each of the upright and supine recovery experiments. Vertical line (at $t = 0$) indicates the start of recovery. Horizontal dashed and dotted lines represent pre-exercise resting levels in upright and supine positions, respectively. Adapted from Takahashi et al. (2000), reproduced by permission

After the transition from the upright to the supine position during recovery after upright exercise, stroke volume increased greatly from the level during upright exercise to the pre-exercise resting supine level (Fig. 2.4a). On the other hand, when the subjects held the same upright position during recovery from upright exercise, stroke volume slightly increased and remained unchanged during a period of about 2 min. Shortly thereafter, both stroke volumes during supine and upright recovery decreased gradually to below the pre-exercise resting levels for the corresponding positions; however, there was no significant difference in resting stroke volume

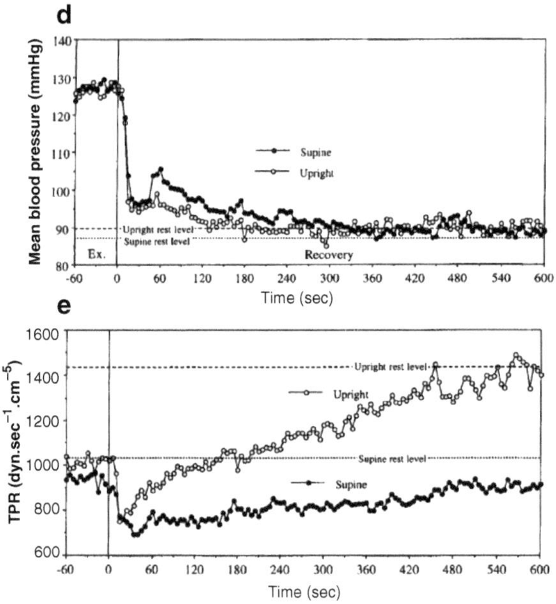

Fig. 2.4 (continued)

between before and after exercise in either position. At the tenth minute of recovery, the mean stroke volume in the upright position was significantly ($P < 0.05$) lower by about 40% than that in the supine position.

Heart rates in both the positions decreased rapidly during the first 120 s of recovery and decreased gradually thereafter (Fig. 2.4b). Importantly, the extent of reduction in supine heart rate was greater than that in upright heart rate. Thus, the mean heart rate at the tenth minute of recovery was significantly lower ($P < 0.05$) in the supine than in the upright position; however, each of them remained significantly ($P < 0.05$) elevated above the pre-exercise level in the corresponding positions (Table 2.1).

Recovery cardiac outputs in the supine and upright positions decreased slowly towards the corresponding baseline levels at rest before exercise (Fig. 2.4c). The mean cardiac output at the tenth minute of recovery was significantly greater ($P < 0.05$) in the supine than in the upright position. The recovery cardiac output in the supine position remained significantly elevated ($P < 0.05$) above the pre-exercise baseline level during a 10-min period. In the transition from upright exercise to supine recovery, cardiac output took a slight and transient tumble during an initial 15-s period, during which the body changed from the upright to supine position after the end of exercise.

Recovery mean blood pressure in the supine and upright positions after upright exercise markedly dropped, transiently rose, and gradually declined to the pre-exercise level in the latter half (Fig. 2.4d). The bottom of the initial drop of mean blood pressure appeared at about 30 s in both the positions during the early phase of recovery. Each of the subsequent weak increases in mean blood pressure in both the positions continued until 60 s while little difference was there in mean blood pressure between the two positions. The recovery mean blood pressures for both positions reached almost the same levels in 5 min.

Total peripheral resistances in the supine and upright positions largely decreased immediately after exercise and then gradually increased to the corresponding pre-exercise levels (Fig. 2.4e). The bottom of the transient fall of total peripheral resistance appeared at 15 s in the upright position and at 35 s in the supine position. The mean total peripheral resistance at the tenth minute of recovery was significantly lower ($P < 0.05$) in the supine than in the upright position (Table 2.1).

The amplitude of the high-frequency component of R-R interval variability at rest before exercise was significantly greater ($P < 0.05$) in the supine than in the upright position (Fig. 2.5). Similarly, the HF amplitude of R-R interval variability in six of seven subjects was greater in the supine than in the upright position during the last 2 min of recovery; however, the difference in this case ($n = 7$) was not statistically significant. For each body position, the HF amplitude during recovery was consistently less than that at rest before exercise. These differences were statistically significant ($P < 0.05$) in both the supine and upright positions.

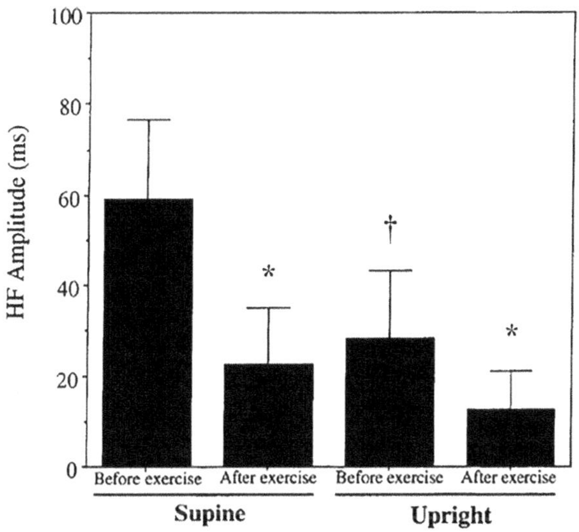

Fig. 2.5 Amplitude of high-frequency (*HF*, 0.15–0.40 Hz) component of R-R interval variability during quiet rest before exercise and during recovery after exercise. Data are group mean values obtained from seven subjects in each of the two experiments with the same upright exercise in regard to the different body position (upright vs. supine) during recovery. $^*P < 0.05$, significantly different from rest before exercise; $^{\dagger}P < 0.05$, significantly different from supine position. Adapted from Takahashi et al. (2000), reproduced by permission

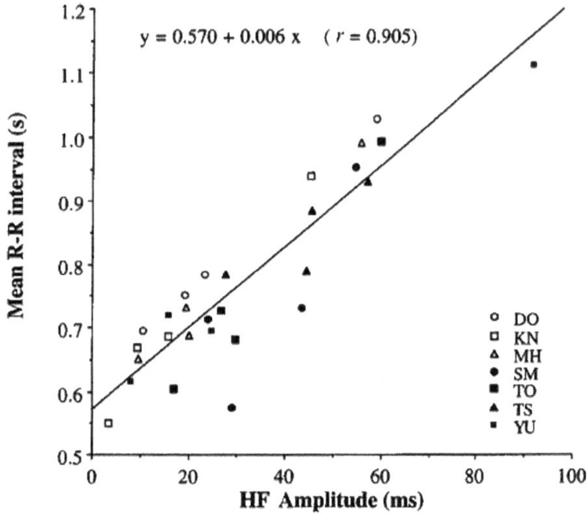

Fig. 2.6 The correlation of high-frequency (*HF*, 0.15–0.40 Hz) amplitude and mean R-R interval for upright and supine positions during rest before exercise and during recovery after exercise. The regression line between the two variables is described as: mean R-R interval = 0.006 × *HF* amplitude +0.570; ($r = 0.905$, $n = 28$, $P < 0.05$). Each symbol represents one subject. Adapted from Takahashi et al. (2000), reproduced by permission

The subjects did not intentionally try to control either their regular breathing rate or its depth, but breathed freely in the experiments. The spontaneous breathing rates in individuals ranged from 9 to 22 breaths per minute in both the pre-exercise and postexercise resting states. The rates were in a range of 0.15–0.37 Hz, which were equivalent to the high-frequency component (*HF*, 0.15–0.40 Hz) in the power spectral analysis.

Figure 2.6 represents a linear relationship between mean R-R interval and HF amplitude at rest before and after exercise in the two positions together for individuals. The linear correlation was statistically significant ($P < 0.05$) between mean R-R interval and *HF* amplitude. The correlation was mathematically described by an equation: *mean R-R interval* = 0.006 × *HF amplitude* + 0.570 ($r = 0.905$, $n = 28$, $P < 0.05$).

2.4 Discussion

2.4.1 Postexercise Tachycardia

It is well established that at rest a decrease in heart rate during postural change from an upright to a supine position is caused by a decrease in cardiac sympathetic nerve activity and an increase in cardiac parasympathetic nerve activity. Similarly, if such reciprocal changes in autonomic nerves to the heart with postural change occur during recovery from exercise, the magnitude of heart rate recovery should be greater in the supine than in the upright position despite almost the same level at the end of the upright exercise. This is the case in this study, in which we have found that the time course of postexercise heart rate after its initial rapid reduction was lower in the supine than in the upright position. Although postexercise heart rate decreased largely during the early phase of recovery, supine heart rate at the tenth minute of recovery remained elevated by about 19 beats/min above the pre-exercise level at supine rest. Similarly, upright recovery heart rate was greater by about 13 beats/min than that of the pre-exercise level at upright rest.

2.4.2 Heart Rate and Cardiac Vagal Activity

As mentioned above, the relatively stable postexercise heart rates after their early reductions in the upright and supine positions remained elevated above the pre-exercise baseline levels. These results are in agreement with those of other studies. Curiously, it seems that though postexercise heart rates are at such a high level, little attention is paid to this elevation as with any important physiological phenomenon. The mechanisms involved in the postexercise heart rate elevation, termed postexercise tachycardia, in both the supine and upright positions have not yet been well understood. In the present study with power spectral analysis, the magnitude of the high-frequency component of R-R interval variability during rest and postexercise recovery was shown to be significantly correlated with the R-R interval, regardless of body position. Such a positive correlation suggests that the heart rate levels are inversely proportional to the cardiac vagal efferent activity. The high-frequency component during recovery from exercise remained reduced below the pre-exercise resting levels in both the upright and supine positions. We also demonstrated that the high-frequency component during recovery as well as that at rest was much greater in the supine than in the upright position. Importantly, the high correlation between the high-frequency amplitude and the mean R-R interval was observed consistently during rest and recovery in the two positions; that is, the lower the high-frequency amplitude, the shorter the R-R interval or the higher the heart rate. Therefore, these results suggest that the delay in the restoration of the cardiac parasympathetic nervous system is related to the prolonged postexercise elevation of heart rate above the pre-exercise baseline levels in both positions.

2.4.3 Transition of Baroreflex-Operating Point

The slow decrease in heart rate after exercise may be partly due to depression of the vagal nerve activity because of a delay in the change of the set point of arterial baroreflex control. The operating point of the baroreflex function shifts to a higher pressure during exercise, which begins at the start of the exercise. The degree of movement of this working point for reflex arterial pressure-heart rate curves will be modestly proportional to the intensity of the exercise or the degree of effort for the exercise. The shift to a higher set point can be explained by the fact that an increase in heart rate is accompanied by an increase in mean blood pressure; simply, resetting the reflex causes the reaction of the two variables to shift in the same directions of increase and decrease. Importantly, we should be aware that vagal activity is normally suppressed under the conditions of physical activity or exercise. After exercise, the transition from the exercising set point of baroreflex to the resting level should occur in the phase of recovery. Since mean blood pressure and heart rate decreased simultaneously in an early phase of the recovery period, it is considered that the operating point of the baroreflex is moving toward the baseline level, though not completely, during the first 2 min of recovery.

Arterial blood pressure returned to the resting level in 5 min of recovery in both upright and supine positions. From this pressure change, it may be reasonably assumed that the stimulated cardiovascular system has returned to the pre-exercise resting level; however, stoke volume, heart rate, cardiac output, and total peripheral resistance did not completely return to resting levels. Thus, we should usually pay attention to the cardiovascular variables other than blood pressure so that postexercise persons, especially those susceptible, can be managed so as to avoid the risk of cardiac arrest or arrhythmia.

Mean blood pressures returned to the pre-exercise resting levels in 5 min of recovery, whereas heart rates were still higher by about 20–30% above resting levels in both the postures. After the period, the heart rate would decrease to the resting baseline level very slowly, for example, over several hours. In this case, the baroreflex function will require the operating point to fall toward a lower level on the time course parallel to a slow decline in heart rate. If the baroreflex set point were not modified, a decrease in heart rate would be prevented by reflexive function. Therefore, delayed resetting of the set point, at which baroreflex works, may also be the cause of delayed recovery of heart rate.

The shift in the operating point, at which the baroreflex efficiently functions, would also be modulated to some extent by nerve inputs relating to body core temperature, active muscle temperature, and metabolic products accumulated in the tissue of the active muscles. These nerve afferents to the central nervous system are likely to take some time to return to their resting levels.

2.4.4 Inhibition of Neuropeptide Y on Vagal Nerve Activity

During an increase in sympathetic nerve activity with exercise, a neural transmitter, norepinephrine, is accompanied by the release of a co-transmitter, neuropeptide Y. Neuropeptide Y is stored in sympathetic nerve terminals throughout the heart. These substances have direct and indirect effects on cardiac functions, such as cardiac cycle length, atrial contractility, and atrioventricular nodal conduction. In a study with electrical stimulation of the stellate ganglia in anesthetized dogs, cardiac sympathetic nerve fibers evoked the inhibitory effect of the vagus nerve on heart rate by the action of neuropeptide Y released from sympathetic nerve terminals. Both the magnitude and duration of the inhibitory effect of the vagus nerve on heart rate were significantly enhanced by an increase in the frequency or duration of sympathetic stimulation. The attenuation of the action of vagus nerves on cardiac pacemaker cells of the sinoatrial node was caused by durable neuropeptide Y for a relatively long period of time. At the time, the effect of the vagal efferents on heart rate gradually recovered and took a few hours to return to a normal level before the sympathetic stimulation. Parenthetically, the neuropeptide Y-induced inhibition of vagal activity on heart rate may result from a chemical reaction of neuropeptide Y, which attenuates acetylcholine release from cardiac vagal nerve endings. In this regard, several sympathetic co-transmitters, other neuropeptides, released from the sympathetic nerves during the antecedent exercise may attenuate vagal action at the heart during recovery. It is also conceivable that the sinoatrial node of the heart is hypersensitive to several inputs from neural signals of the cardiovascular center or to the surrounding physiological conditions including increases in tissue temperature and accumulated metabolites.

2.4.5 Cardiopulmonary Mechanoreflex

When the body posture changes from a supine to an upright position, blood in the thoracic vascular compartment is translocated into the lower back, buttocks and pelvis, and the lower extremities, thereby leading to falls in both the ventricular filling and stroke volume. Conversely, in the present study, when the subjects moved from the upright sitting position to the supine lying position after exercise, stroke volume increased markedly within 90 s after the posture change during recovery. This increase in stroke volume would be closely associated with an increase in central venous blood volume (the preload) and a decrease in total peripheral resistance (the afterload). If such an increase in central venous volume is induced during supine recovery, the sympathetic vasoconstrictor nerves innervating resistance vessels of splanchnic and skeletal muscle vascular beds may be reduced by the tonic inhibition of cardiopulmonary mechanoreceptor reflex, compared with upright recovery.

On the other hand, a sudden, transient decrease in total peripheral resistance immediately after exercise results from the release of vascular compression due to the tension developed by the working muscle fibers; that is, the short-term reduction in resistance is due to the mechanical factor outside the blood vessels. Similar transitory reductions in total peripheral resistance that appeared in sitting and lying positions provide evidence that minimal neural reflex by cardiopulmonary receptors is responsible for these prompt changes in total peripheral resistance.

As mentioned above, postexercise sympathetic nerve activity in the supine position is suppressed reflexly by the cardiopulmonary mechanoreceptors, which are loaded by increased blood volume in the central venous system translocated from the lower extremities. This neural reflex of the cardiopulmonary receptors may be explained by the fact that the recovery of total peripheral resistance to the pre-exercise resting level is delayed in the supine position compared with the sitting position. Apart from the sudden decrease in total peripheral resistance, the relationship between the subsequent slow responses of total peripheral resistance and possible sympathetic nerves is consistent with the tonic inhibition of cardiopulmonary nerve traffic due to the redistribution of blood into the central venous system in supine posture during the recovery period.

2.4.6 Adjustments of Blood Pressure

Immediately after the end of the exercise, the initial sudden falls in mean blood pressure observed in both the positions were induced mainly by the large reductions in total peripheral resistance, since the reductions in cardiac output in upright and supine positions were more gradual. Although actual values of cardiac output and total peripheral resistance were largely different between the upright and supine positions, mean blood pressures were approximately equal between the two positions after 5 min of recovery. Thus, these findings suggest that neurohumoral control mechanisms are involved in the different responses of heart rate, cardiac output, and total peripheral resistance to recovery in the upright and supine positions. As already described, the cardiovascular control center in the central nervous system is likely to cause the operating point of baroreflexes to adjust the systemic arterial blood pressure. This adjusted blood pressure should provide adequate muscle perfusion for the washout of metabolic substances as also refresh the nutrition within the postexercise muscular tissues.

2.4.7 Blood Pressure Regulation During Posture Changes

The hypotensive effect of exercise on arterial blood pressure in hypertensive and normotensive subjects has been investigated extensively during recovery from dynamic exercise. The postexercise hypotension can be maintained by differently regulating the cardiovascular variables dependent on body positions. It has been proposed so far that the arterial blood pressure of the systemic circulation during postural changes is maintained by fine adjustments of peripheral vascular resistance and cardiac output. The magnitude of adjustment of both variables follows the different modes of circulatory control in body postures. There are substantial differences in the reflex nerve afferents of the cardiopulmonary, carotid sinus, aortic arch mechanoreceptors, and the muscle chemoreceptors to the central nervous system between the upright and supine positions. If one study is conducted on the circulatory control mechanisms in conjunction with the upright position alone, it may mislead us in assessing variables that play relative roles in regulating arterial blood pressure during recovery from exercise. Therefore, it is important to understand the roles of the variables in the frame of different postural context, such as upright and supine positions.

For the mechanisms of postexercise hypotension, we must consider relative roles of individual stretch sensors in controlling arterial and venous circulation. In the systemic circulation under different physical situations, some classes of receptors were simultaneously affected by alterations in body posture. For example, in supine posture, the hydrostatic pressure difference between the carotid sinus baroreceptors and the aortic arch baroreceptors was about 0 mmHg. In contrast, in upright posture, the intravascular pressure at the carotid baroreceptors was 15–18 mmHg lower than that at the aortic baroreceptors, since there is a distance of 20–25 cm between the two receptors. The hydrostatic pressure difference between the two receptor sites provides the central nervous system with information about the gravitational effect, i.e., the tilt angle of the body axis.

In orthostasis, also, with a supplement of vestibular afferent information, the arterial baroreceptors and cardiopulmonary mechanoreceptors will all change their afferent discharge in proportion to the degree of posture tilt. All of these various afferents to the central nervous system would evoke changes in total peripheral resistance and cardiac output via the efferents of integrated neuronal signals, primarily increased by the sympathetic nerve activity. The sympathetic nerve outflow to the heart and peripheral vasculature is coordinated to adjust the cardiovascular response necessary for maintaining homeostasis of fluid distribution in the whole body. To achieve this homeostasis, cardiac output and total peripheral resistance must be regulated to maintain blood pressure, and the blood pressure level at the time provides the two regulators with crucial information about a target point of blood pressure via the neural feedback mechanisms. Changes in both sympathetic vasoconstrictor drive and local metabolic rate would reset autoregulation and baroreflex control to operate at a new appropriate level for blood pressure and flow in inactive and active organs. In this context, it should be emphasized that the regulators of the circulatory system for postexercise hypotension are not always identical in the relative contributions to blood pressure control in different postures.

2.4.8 *Baroreflex Pathways*

There is an essential spot of first synapse for all the cardiovascular afferents from arterial baroreceptors, arterial chemoreceptors, cardiopulmonary receptors, and somatic receptors; that is, the nucleus tractus solitarius (NTS) is located in the dorsomedial medulla (Fig. 2.7). The aortic arch and carotid sinus baroreceptors are

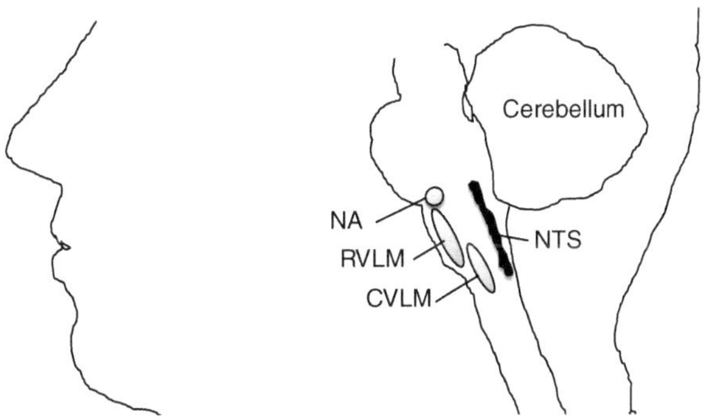

Fig. 2.7 Important areas involved in baroreflexes. *NA* nucleus ambiguus, *NTS* nucleus tractus solitarius, *RVLM* rostral ventrolateral medulla, *CVLM* caudal ventrolateral medulla

stimulated by distention of the vessel wall in which they are located. The neurons of these afferent endings discharge at an increased rate when the pressures in these vessels rise. Their afferent nerve fibers pass via the glossopharyngeal and vagus nerves to the medulla oblongata. Most of these fibers terminate in the nucleus tractus solitarius, and then the excitatory transmitter they secrete is glutamate. Excitatory projections from the nucleus tractus solitarius reach the caudal ventrolateral medulla, where they stimulate γ-aminobutyrate-secreting inhibitory neurons that project to the rostral ventrolateral medulla. The rostral ventrolateral medulla is one of the major sources of excitatory input to sympathetic nerves controlling the vasculature. Excitatory glutamate projections also extend from the nucleus tracuts solitarius to the vagal motor neurons in the nucleus ambiguous and the dorsal motor nucleus. Thus, increased baroreceptor traffic excites the vagal outflow to the heart and reduces the tonic sympathetic outflow to the heart and peripheral vessels. When blood pressure increases, these neural pathways reflexly produce reductions in peripheral vascular resistance, heart rate, and cardiac output. As a result, the arterial pressure will return to its original-set level.

2.4.9 Group III and IV Afferent Reflexes

Impulses generated in chemical and mechanical sensor endings in working muscle tissues are transmitted to afferent neurons that have axons terminating in the central nervous system, particularly in the nucleus tractus solitarius in the medulla. The somatic afferent fibers carrying the excitatory input from skeletal muscles are small, myelinated group III and small, unmyelinated group IV or C fibers. Group III fibers are chiefly sensitive to mechanical stretch and group IV fibers are chiefly sensitive to chemicals released during exercise, notably potassium ions and hydrogen ions (due to lactic acid formation). Importantly, group III and IV afferent fibers are also thermosensitive; particularly, increased muscle temperature in a range of 24–44 °C may elevate the firing rates of both groups III and IV afferents.

2.4.10 Interaction of Baroreflexes

The blood pressure responses to exercise are caused both by the increases in heart rate and cardiac output and by the vasoconstrictions in non-exercising muscles and splanchnic vascular beds. Activation of arterial baroreceptors during exercise cannot concomitantly inhibit the excitatory sympathetic drive originating from the central command or from the somatic reflexes. However, the arterial baroreceptors still modulate changes in arterial pressure and heart rate when disproportionated fluctuations in blood pressure appear. Various studies in exercising humans have indicated that in general the regulation of arterial pressure and heart rate by baroreceptors is essentially unchanged during exercise and recovery, except for the reflex bradycardia.

Apparently, the excitatory influence of arterial baroreceptors on vagal cardiac inhibitory neurons causing bradycardia may be partly opposed during exercise, possibly because of hyperventilation or interaction with central suprabulbar pathways. Other possible variations in the set point of baroreflexes are also associated with responsiveness of the sinus node, alertness and sedation levels, and respiration. A detailed interpretation of these factors involved in their interactions would characterize baroreflex control of the central nervous system.

2.4.11 Limitations

Average values for cardiac output measured by impedance cardiography during exercise in the present study were approximately 30% lower than those obtained by other techniques at a similar relative intensity of exercise based on maximal oxygen consumption, albeit with a different average body size. The absolute values for impedance-derived cardiac output remain to be elucidated. However, there has been a general agreement that impedance cardiography is an acceptable way of estimating relative changes in cardiac output during exercise and during postural changes. Furthermore, in this study, we have verified that the application of impedance cardiography can be useful to measure stroke volume in postural change during the period of recovery from exercise. The proof of this verification is that the time courses of relative changes in stroke volume in both the upright and supine positions are consistently similar to those obtained by pulsed Doppler ultrasoundgraphy in a study by Johnson et al. The main interest of this study is relative changes in measurement rather than absolute values. Therefore, some errors that may be involved in impedance-derived cardiac output do not seriously affect the results.

The central venous pressure is known to increase slightly from the upright to the supine position. In the present study, however, total peripheral resistance was estimated on the assumption that the central venous pressure in the subjects was close to 0 mmHg throughout the experiments. The computed total peripheral resistance in the supine position may thus be overestimated by up to 15% at most; however, this has little serious effect on our conclusions.

Both the rate and depth of breathing must usually be controlled to estimate cardiac vagal nerve activity through the high-frequency component of heart rate variability on power spectral analysis. In this study, subjects did not always breathe in the same pattern, but breathed spontaneously; however, mean values for respiratory frequency were not significantly different between the pre-exercise rest and postexercise recovery or between the upright and supine positions. There may be somewhat ambiguous aspects of the coupling of high-frequency and mean R-R interval, which are mediated by vagal nerve activity. On the contrary, if breathing is controlled, for example, a constant respiratory frequency, it may raise the problem that maintenance of respiratory homeostasis may not work well in the recovery period after exercise. It has been noted in the published literature that both respiratory rhythm and pattern are different between interindividual and intraindividual subjects at rest both before and after exercise.

It is clear that activity in cardiac sympathetic efferent nerves contributes to the low-frequency component (0.04–0.15 Hz) of heart rate variability. The low-frequency component is also markedly affected by activity in cardiac vagal efferent nerves. Moreover, it has not been well documented how the postural changes affect the interaction between the parasympathetic and sympathetic afferent nerves or the relative contribution of vagal and sympathetic mechanisms to heart rate control during recovery, as compared with those at rest. Thus, it is difficult to interpret both cardiac sympathetic nerve activity and sympathovagal balance in terms of the low-frequency component and the ratio of low- to high-frequency component during recovery from exercise. Clearly, additional research is necessary to clarify the interaction between the parasympathetic and sympathetic nerve traffic to the pacemaker cells of the heart during recovery from exercise.

2.5 Summary

Cardiovascular responses were examined in seven healthy male subjects during 10 min of recovery in the upright and supine positions after 5 min of upright cycle exercise at 80% of peak oxygen uptake. Heart rate rapidly decreased during the early phase of recovery and then decreased much slowly in both the upright and supine positions. The mean heart rate at the tenth minute of recovery was significantly lower in the supine than in the upright position. The heart rate at the tenth minute was, however, significantly higher than the pre-exercise resting level in each position. Accordingly, the amplitude of the high-frequency (*HF*) component of R-R interval variability in power spectral analysis was less with decreased mean R-R interval during upright recovery than during supine recovery. The *HF* amplitude and mean R-R interval during recovery were also significantly less than those at rest before exercise in each position. The close relationship between the two variables was expressed by a regression line: *mean R-R interval* = 0.006 × *HF amplitude* + 0.570 ($r = 0.905$, $n = 28$, $P < 0.05$) for data together with both before and after exercise in the two positions. These results suggest that after the early phase of recovery, the heart rate elevated above the pre-exercise resting level is associated with the retarded restoration of the cardiac parasympathetic nervous system, regardless of body position. Heart rate and total peripheral vascular resistance were significantly higher during upright recovery. In contrast, impedance-derived stroke volume and cardiac output were significantly lower during upright recovery. However, mean arterial blood pressures in response to recovery were approximately the same in the upright and supine positions. Therefore, the similar recovery of blood pressure is exquisitely achieved through the cardiovascular regulation of heart rate, stroke volume, cardiac output, and total peripheral resistance in a reciprocal manner to offset each other between the upright and supine positions.

Further Reading

Takahashi T, Okada A, Saitoh T, Hayano J, Miyamoto Y. Difference in human cardiovascular response between upright and supine recovery from upright cycle exercise. Eur J Appl Physiol. 2000;81:233–9.

Arai Y, Saul JP, Albrecht P, Hartley LH, Lilly LS, Cohen RJ, Colucci WS. Modulation of cardiac autonomic activity during and immediately after exercise. Am J Phys. 1989;256:H132–41.

Bjurstedt H, Rosenhamer G, Balldin U, Katkov V. Orthostatic reactions during recovery from exhaustive exercise of short duration. Acta Physiol Scand 1983;119:25–31.

Brown SP, Li H, Chitwood LF, Anderson ER, Boatwright D. Blood pressure, hemodynamic, and thermal responses after cycling exercise. J Appl Physiol. 1993;75:240–5.

Casadei B, Cochrane S, Johnston J, Conway J, Sleight P. Pitfalls in the interpretation of spectral analysis of the heart rate variability during exercise in humans. Acta Physiol Scand. 1995;153:125–31.

Cerretelli P, di Prampero PE. Gas exchange in exercise. In: Farhi LE, Tenney SM, editors. Handbook of physiology. The respiratory system. Bethesda, MD., sect 3, vol. 1, chap 16: American Physiological Society; 1987. p. 297–339.

Cléroux J, Kouamé N, Nadeau A, Coulombe D, Lacourcière Y. After effects of exercise on regional and systematic hemodynamics in hypertension. Hypertension. 1992;19:183–91.

Dampney RAL. Functional organization of central pathways regulating the cardiovascular system. Physiol Rev. 1994;74:323–64.

Denniston JC, Maher JT, Reeves JT, Cruz JC, Cymerman A, Grover RF. Measurement of cardiac output by electrical impedance at rest and during exercise. J Appl Physiol. 1976;44:91–5.

Eckberg DL. Sympathovagal balance. A critical appraisal. Circulation. 1997;96:3224–32.

Furlan R, Piazza S, Dell'Orto S, Gentile E, Cerutti S, Pagani M, Malliani A. Early and late effects of exercise and athletic training on neural mechanisms controlling heart rate. Cardiovasc Res. 1993;27:482–8.

Gauer OH, Thron HL. Postural changes in the circulation. In: Hamilton WF, Dow P, editors. Handbook of physiology. Circulation. Washington DC., sect 2, vol 3, chap 67: American Physiological Society; 1965. p. 2409–39.

Hagberg JM, Montain SJ, Martin WH III. Blood pressure and hemodynamic responses after exercise in older hypertensives. J Appl Physiol. 1987;63:270–6.

Halliwill JR, Taylor JA, Eckberg DL. Impaired sympathetic vascular regulation in humans after acute dynamic exercise. J Physiol. 1996;495:279–88.

Hayano J, Mukai S, Sakakibara M, Okada A, Takata K, Fujinami Y. Effects of respiratory interval on vagal modulation of heart rate. Am J Phys. 1994;267:H33–40.

Heymans C, Neil E. Reflexogenic areas of the cardiovascular system. London: J & A Churchill Ltd.; 1958.

Imholz BPM, Settels JJ, Van Der Meiracker AH, Wesseling KH, Wieling W. Non-invasive continuous finger blood pressure measurement during orthostatic stress compared to intra-arterial pressure. Cardiovasc Res. 1990;24:214–21.

Johnson EC, Hudson TL, Greene ER. Left ventricular hemodynamics during exercise recovery. J Appl Physiol. 1990;69:104–11.

Komaru T, Ashikawa K, Kanatsuka H, Sekiguchi N, Suzuki T, Takishima T. Neuropeptide Y modulates vasoconstriction in coronary microvessels in the beating canine heart. Circ Res. 1990;67:1142–51.

Kubicek WG, Karnegis JM, Patterson RP, Witsoe DA, Mattson RH. Development and evaluation of an impedance cardiac output system. Aerosp Med. 1966;37:1208–12.

Mark AL, Mancia G. Cardiopulmonary baroreflexes in humans. In: Shepherd JT, Abboud FM, Geiger SR, editors. Handbook of physiology. The cardiovascular system. Peripheral circulation and organ blood flow. Bethesda, MD., sect 2, vol 3, part 3, chap 21: American Physiological Society; 1983. p. 795–813.

Mitchell JH, Schmidt RF. Cardiovascular reflex control by afferent fibers from skeletal muscle receptors. In: Shepherd JT, Abboud FM, Geiger SR, editors. Handbook of physiology. Peripheral circulation and organ blood flow. Bethesda, MD., sect 2, vol 3, Chap 17: American Physiology Society; 1983. p. 623–58.

Miyamoto Y, Hiura T, Tamura T, Nakamura T, Higuchi J, Mikami T. Dynamics of cardiac, respiratory, and metabolic function in men in response to step work load. J Appl Physiol. 1982a;52:1198–208.

Miyamoto Y, Tamura T, Hiura T, Nakamura T, Higuchi J, Mikami T. The dynamic response of the cardiopulmonary parameters to passive head-up tilt. Jpn J Physiol. 1982b;32:245–58.

Miyamoto Y, Higuchi J, Abe Y, Hiura T, Nakazono Y, Mikami T. Dynamics of cardiac output and systolic time intervals in supine and upright exercise. J Appl Physiol. 1983;55:1674–81.

Mukai S, Hayano J. Heart rate and blood pressure variabilities during graded head-up tilt. J Appl Physiol. 1995;78:212–6.

Oka S. Cardiovascular Hemorheology. Cambridge: Cambridge University Press; 1981.

Perini R, Orizio C, Comandè A, Castellano M, Beschi M, Veicsteinas A. Plasma norepinephrine and heart rate dynamics during recovery from submaximal exercise in man. Eur J Appl Physiol. 1989;58:879–83.

Pomeranz B, Macaulay RJB, Caudill MA, Kutz I, Adam D, Gordon D, Kilborn KM, Barger AC, Shannon DC, Cohen RJ, Benson H. Assessment of autonomic function in humans by heart rate spectral analysis. Am J Phys. 1985;248:H151–3.

Potter EK. Prolonged non-adrenergic inhibition of cardiac vagal action following sympathetic stimulation: neuromodulation by neuropeptide Y. Neurosci Lett. 1985;54:117–1021.

Potter EK. Presynaptic inhibition of cardiac vagal postganglionic nerves by neuropeptide Y. Neurosci Lett. 1987;83:101–6.

Potter EK. Effects of neuropeptides on the vagal neuroeffector junction. In: Levy MN, Schwartz PJ, editors. Vagal control of the heart: experimental basis and clinical implications. New York: Futura; 1994. p. 289–303.

Robinson BF, Epstein SE, Beiser GD, Braunwald E. Control of heart rate by the autonomic nervous system. Studies in man on the interrelation between baroreceptor mechanisms and exercise. Circ Res. 1966;19:400–11.

Rowell LB. Reflex control during orthostasis. In: Rowell LB, editor. Human Cardiovascular Control. New York: Oxford University Press; 1993. p. 37–80.

Sawada Y, Ohtomo N, Tanaka Y, Tanaka G, Yamakoshi K, Terachi S, Shimamoto K, Nakagawa M, Satoh S, Kuroda S, Imurai O. New technique for time series analysis combining the maximum entropy method and non-linear least squares method: its value in heart rate variability analysis. Med Biol Eng Comput. 1997;35:318–22.

Smith JJ, Ebert TJ. General response to orthostatic stress. In: Smith JJ, Ebert TJ, editors. Circulatory response to the upright posture. Boca Raton, FL: CRC Press; 1990. p. 1–46.

Somers VK, Abboud FM. Baroreflexes in health and disease. In: Levy MN, Schwartz PJ, editors. Vagal control of the heart: experimental basis and clinical implications. New York: Futura; 1994. p. 381–401.

Takahashi T, Miyamoto Y. Influence of light physical activity on cardiac responses during recovery from exercise in humans. Eur J Appl Physiol. 1998;77:305–11.

Takahashi T, Niizeki K, Miyamoto Y. VE-VCO_2 relationship in transient responses to step-load exercise from rest to recovery. In: Hughson RL, Cunningham DA, Duffin J, editors. Advances in modeling and control of ventilation. New York: Plenum Press; 1998. p. 207–12.

Takahashi T, Yamamoto S, Nishimura H, Watanabe S, Mimura Y, Saitoh T, Sugimoto T, Yanagida H, Yoshimura A. An allometric scaling mathematical model for capillary-covered alveoli in mammalian lungs. Ther Res. 2023;44:295–302.

Task Force for The European Society of Cardiology and The North American Society of Pacing and Electrophysiology. Heart rate variability. Standards of measurement, physiological interpretation, and clinical use. Circulation. 1996;93:1043–65.

Warner MR, Levy MN. Neuropeptide Y as a putative modulator of the vagal effects on heart rate. Circ Res. 1989;64:882–9.
Warner MR, Senanayake P, Ferrario CM, Levy MN. Sympathetic stimulation-evoked overflow of norepinephrine and neuropeptide Y from the heart. Circ Res. 1991;69:455–65.
Watson RDS, Hamilton CA, Jones DH, Reid JL, Stallard TJ, Littler WA. Sequential changes in plasma noradrenaline during bicycle exercise. Clin Sci. 1980;58:37–43.

Chapter 3
Vagal Activity and Duration of Recovery Exercise

3.1 Introduction

Sudden cardiac death in sports may occur not only during physical activity but also during recovery from exercise. An extended cool-down exercise or active recovery after exercise is widely believed to play important roles in protection against cardiac events and in removal of accumulation of metabolic wastes within the circulation and active muscular tissues. A large number of studies have shown that the removal of fatigue-related substances, such as lactic acid, produced during strenuous exercise is markedly enhanced in the circulation by light to moderate exercise during the period of recovery. Thus, cool-down exercise during recovery appears to provide a facilitatory effect on restoring cardiovascular and respiratory systems from the metabolic perturbations due to prior intense exercise.

It has long been known that immediately after exercise, heart rate decreases rapidly during an early phase of recovery and decreases more slowly to a relatively stable level during the subsequent phase. This stable level of recovery heart rate remains elevated above its basal pre-exercise level for a prolonged period. When a person takes a complete rest without cool-down exercise, the elevation in postexercise heart rate is sustained far beyond a time that is required to return respiratory gas exchange to its resting level. Some pieces of our research have demonstrated that a protocol of cool-down exercise during recovery effectively reduces the time it takes to recover heart rate to the pre-exercise resting level, when compared with a protocol without cool-down exercise (see Chap. 1).

To date, it has been shown that cool-down exercise during recovery from exercise plays important roles both in facilitating the cardiac filling through the venous return of blood to the heart and in preventing the accumulation of blood in the postexercise extremities in an upright position by the pumping action of the skeletal

T. Takahashi, *The Cardiovascular System During Exercise and Recovery*, https://doi.org/10.1007/978-981-99-8486-2_3

muscles, the so-called skeletal muscle pump. Several researchers have advocated that cool-down exercise or active recovery exercise has a physiological effect on suppressing cardiac arrhythmia and cardiac arrest, namely cardiac catastrophes. In the literature, it has also been suggested that cool-down exercise with a gradually decreasing work load can reduce the risk of cardiac events; however, there has been little evidence for a good effect of cool-down exercise on cardiac functions, such as chronotropic action (contraction rate of the heart) and inotropic action (contractility of the heart). Moreover, few attempts have been made to ascertain whether the recovery of the cardiovascular system from exercise is affected by the difference in duration of cool-down exercise. Thus, knowledge regarding the effects of duration of cool-down exercise on cardiac recovery has hitherto been limited.

Some human studies have assessed autonomic control of heart rate during recovery from exercise with the use of a noninvasive technique of power spectral analysis of heart rate variability. An exponential deceleration in heart rate after exercise is accompanied by an increase in the amplitude of high-frequency (0.15–0.40 Hz [hertz]) components of heart rate variability and by a decrease in the amplitude of low-frequency components. These results have suggested that an initial fast decrease and its subsequent much slower decrease in heart rate after heavy exercise are mediated by prompt reactivation of the parasympathetic nervous system and gradual withdrawal of the sympathetic nervous system. However, it has not been well understood whether the recovery of postexercise heart rate enhanced by additional active-recovery exercise is associated with changes in the two divisions of the autonomic nervous system, particularly in cardiac vagal efferent nerve activity. The reason for our focus on vagal nerve activity is that vagal outflow to the heart in the resting state is dominant over sympathetic nerve outflow. Also, returning parasympathetic activity to the basal resting level is of great interest because vagus nerve tone is expected to suppress sudden cardiac death after exercise.

Therefore, the first part of this chapter characterizes the effects of different lengths of cool-down exercise on heart rate during recovery from heavy exercise, which is dealt with in Experiment 1. The second part also explores the relationship between cardiac vagal nerve activity and heart rate, with or without cool-down exercise, returning forward the pre-exercise resting level during recovery from exercise, which is dealt with in Experiment 2. The research hypothesis is that the vagal efferent is evoked with the heart rate recovery by cool-down exercise.

The study of Experiment 2 analyzes a relationship between the high-frequency amplitude of heart rate variability and the fluctuations of beat-to-beat heart rate. The high-frequency amplitude, as an index of cardiac vagal nerve activity, is assessed by power spectral analysis in subjects, who perform the same exercise with the following two different recovery modes: (1) quiet rest without cool-down exercise (inactive recovery), and (2) quiet rest after a light intensity cool-down exercise (partially active recovery).

3.2 Materials and Methods

3.2.1 *Human Subjects*

3.2.1.1 Experiment 1

Seven healthy untrained male subjects participated in this study. The physical characteristics of the participants with a mean age of 22.4 ± 2.1 years (mean ± SD) were as follows: a mean height, 1.70 ± 0.06 m; a mean body mass, 61.1 ± 9.1 kg; and a mean peak oxygen uptake, 2.73 ± 0.40 L/min.

3.2.1.2 Experiment 2

Six healthy young male subjects participated in this study. The physical characteristics of the participants with a mean age of 22.7 ± 2.3 years (mean ± SD) were as follows: a mean height, 1.71 ± 0.06 m; a mean body mass, 61.8 ± 9.7 kg; and a mean peak oxygen uptake, 2.64 ± 0.36 L/min.

For determination of peak oxygen uptake, each participant performed a ramp-exercise test with an incremental work rate of 15 watts every minute until they could no longer continue to exercise on an electromagnetically braked cycle ergometer (Corival 300; Lode, Netherlands).

All the voluntary participants undergoing Experiments 1 and 2 were fully informed about the procedures, risks, and benefits of each study after a medical examination. They provided a written informed consent before the study. None of the participants had any history of circulatory or respiratory disease. On a daily basis, they do not participate in any regular physical activity or intense competitive sports. The two experiments were conducted in accordance with the principles outlined in the Declaration of Helsinki (ethical code of medical research for humans) and with approval from the institutional review board at the university.

3.2.2 *Experimental Protocols*

3.2.2.1 Experiment 1

Each participant completed two repetitions of each test in terms of three different lengths of recovery exercise. The experimental procedures were similar to those described in detail in Chap. 1. The participants rested for 5 min in an upright seated position on the cycle ergometer and then performed a warm-up of unloaded cycling for 3 min. They exercised for 5 min at a constant intensity equivalent to 70% of peak oxygen uptake. The recovery from exercise was evaluated under the following three conditions: (1) complete rest for 10 min (inactive recovery); (2) cycle exercise at 20% of peak oxygen uptake for 3 min followed by complete rest for 7 min (partially active recovery I); and (3) cycling at 20% of peak oxygen uptake for 6 min followed by complete rest for 4 min (partially active recovery II).

3.2.2.2 Experiment 2

Each participant underwent two exercise tests with different recovery modes after the same exercise. Each test consisted of rest for 5 min on the ergometer, cycle exercise for 5 min, and one of two types of recovery for 10 min. The constant intensity of exercise was equivalent to 70% of peak oxygen uptake. After the exercise, the two types of recovery were separately performed in the following different modes: (1) complete rest for 10 min (inactive recovery) and (2) cycle exercise at 20% of peak oxygen uptake for 6 min (cool-down exercise) followed by complete rest for 4 min (partially active recovery). All participants conducted each of the exercise tests twice.

3.2.3 Experimental Conditions

In Experiments 1 and 2, the participants were in a sitting position on the cycle ergometer during the period of the experiments, including before, during, and after exercise. They pedaled at a constant rate of 60 revolutions per minute and maintained the pedaling rate with the aid of an audible metronome. The pedaling rate was checked with an inductive revolution meter fitted to the ergometer. They were instructed to place their feet on the footrest next to the flywheel during the rest periods. At the time, they also avoided any movement of the upper and lower extremities, so as not to enhance the venous return of blood to the heart by the skeletal muscle pump.

Each participant was allowed to breathe freely in the pattern and depth of breathing or rhythm throughout the experiments.

The participants were instructed not to consume caffeine or alcohol at least 24 h before the experiments. They had eaten a light meal at least 3 h before the exercise tests. The exercise tests started at 10:00 a.m. The order of the tests was randomly assigned among the participants. They were scheduled to participate in one of the exercising tests 1 day in a week; the reason for this was that they should not enhance the cardiorespiratory capacity as exercise training.

All experiments were conducted at a room temperature of 19–23 °C (degrees Celsius).

3.2.4 Measurements of Cardiorespiratory Variables

Tidal volume (V_T, in liters), respiratory frequency (*f*, in breaths per minute), and oxygen uptake (VO_2, in liters per minute) were recorded on a breath-by-breath basis with a computerized measurement system. The participants wore a face mask, the front entrance of which had a hot-wire transducer. Inspiratory and expiratory gases could flow bidirectionally through the hot-wire transducer. The electrical signal produced from the bidirectional flow by the transducer was transmitted to a pneumotachograph (RF-2; Minato, Osaka, Japan); thus, air taken into the lungs and out again from the mouth was monitored as ventilatory airflow with this pneumotachograph. A medical mass spectrometer (WSMR-1400; Westron, Chiba, Japan) was used to analyze small samples of ventilatory gasses. The samples were introduced into the spectrometer via a narrow-bore tube containing only a little dead-space volume from a cylindrical outlet of side wall of the face mask. The composition of breaths was continuously analyzed with the medical mass spectrometer. The spectrometer was calibrated with a gas mixture as a standard reference before each measurement.

The electrocardiographic signal was monitored at the chest wall with bipolar electrodes and digitized at a sampling rate of 500 Hz. The temporal positions of R waves were identified by a peak detection algorithm. Heart rate was calculated from R-R intervals. For further details on the measurement methods, see Chap. 2.

3.2.5 Power Spectral Analysis

A maximal entropy method was used to perform power spectral analysis on R-R intervals during a 2-min period from the fifth to sixth minute inclusive and during a 3-min period from the eighth to tenth minute inclusive of recovery. An R-R interval series of 2 min or longer is required to compute stable power spectral density. Linear trends, if any, were removed from the time series of R-R interval before spectral analysis. The amplitude of a high-frequency component (HF, 0.15–0.40 Hz) of R-R interval variability was assessed. The analysis of the maximal entropy method was performed by using a commercially available computerized system (MemCal; Suwa-Torasuto, Tokyo, Japan). The validity of this spectral analysis system has been confirmed for R-R interval data in several human studies.

In this study, the amplitude of the high-frequency component was expressed as the mean amplitude in milliseconds (ms), which was equal to $(2 \times \text{power of component})^{1/2}$. The high-frequency amplitude of R-R interval variability is a good index of vagal nerve outflow to the heart. Although low-frequency components are often used to assess sympathetic nerve activity to the heart, it is still quite difficult to interpret their physiological role in the presence of cardiac vagal efferent nerves. Thus, the analysis of low-frequency components was excluded from this study.

3.2.6 Data and Statistical Analyses

3.2.6.1 Experiment 1

All breath-by-breath data obtained from each participant, who performed the same exercise with three different recovery modes, were regularly interpolated at five-second intervals with the use of Lagrange polynomials. The two interpolated data for the same recovery were then averaged to yield a single set of data per participant. Group mean values were obtained from the seven individual sets of data. The mean values for heart rate were determined from the last minute of averaged data before, during, and after exercise. Differences in mean heart rate at the tenth minute of recovery were assessed with the use of one-way analysis of variance. As a multiple comparison test, two-sided paired *t* test was used to determine the significance of differences.

3.2.6.2 Experiment 2

Six participants underwent two different recovery tests, each one of which was repeated twice a week apart. Cardiorespiratory data of two repetitions for each participant were averaged to yield a single set of data.

The heart rate and the high-frequency amplitude from the fifth through sixth minutes were averaged during both active and inactive recovery. Their differences were analyzed with the use of a Wilcoxon signed-ranks test. Differences in mean value for each variable during the last 3 min (minutes 7 to 10) of recovery after the two different recovery modes were also analyzed with the use of Wilcoxon's signed-rank test.

All data are expressed as mean ± standard deviation (SD). For all statistical analyses, differences were considered significant at the $P < 0.05$ level.

3.3 Results

3.3.1 Experiment 1

Figure 3.1 shows the temporal changes in heart rate during inactive recovery (complete rest) for 10 min, during partially active recovery I (3-min cycling at 20% of peak oxygen uptake plus 7-min rest), and partially active recovery II (6-min cycling at 20% of peak oxygen uptake plus 4-min rest). For inactive recovery experiments, heart rate was 82.3 ± 7.9 beats/min at rest and then 146.0 ± 17.8 beats/min during exercise at a relative intensity of 70% of peak oxygen uptake. Heart rate during inactive recovery decreased rapidly during the first 2 min and decreased gradually thereafter. Similar decreases in heart rate were observed during the first 2 min of

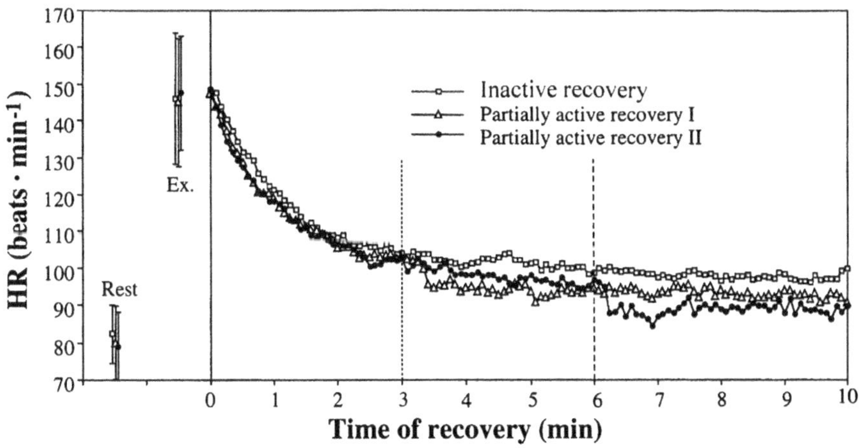

Fig. 3.1 Heart rate (HR) at rest, during exercise (Ex.) at 70% of peak oxygen uptake, and during (1) inactive recovery without cycling for 10 min, (2) partially active recovery I with 3 min of cycling at 20% of peak oxygen uptake plus 7 min of rest, and (3) partially active recovery II with 6 min of cycling at 20% of peak oxygen uptake plus 4 min of rest for seven subjects. The vertical solid line (at $t = 0$) indicates the termination of exercise, the vertical dotted line, the termination of 3-min cycling for partially active recovery I, and the vertical broken line, the termination of 6-min cycling for partially active recovery II. Adapted from Takahashi et al. 2000a, reproduced by permission

two partially active recovery modes. However, after the two active recovery periods of 3-min and 6-min cycling, each heart rate fell conspicuously. These decreased heart rates at rest after active recovery were lower than the heart rate during inactive recovery without low-intensity cycling.

Figure 3.2 shows mean heart rates for individuals obtained during the last tenth minute of recovery. On an individual basis, the resting heart rate at the tenth minute was significantly ($P < 0.05$) less after cessation of both the active recovery modes than after inactive recovery. Although there were interindividual variations in the degree of decreased heart rate, heart rate tended to decrease depending on the length of time, during which cool-down exercise of cycling was performed, except in one participant (TS).

3.3.2 *Experiment 2*

Figure 3.3a shows dynamic changes in heart rate with beat-to-beat fluctuations during the last minute of exercise at 70% of peak oxygen uptake and during two modes of recovery with and without 6 min of cool-down exercise at 20% of peak oxygen uptake both followed by a quiet rest of 4 min in a representative participant. The mean heart rates at the fifth minute of high-intensity exercise were almost the same before the recovery of inactive (complete rest) and active (cool-down exercise) modes: 149.7 ± 17.1 and 148.7 ± 16.9 beats/min, respectively. After cessation of

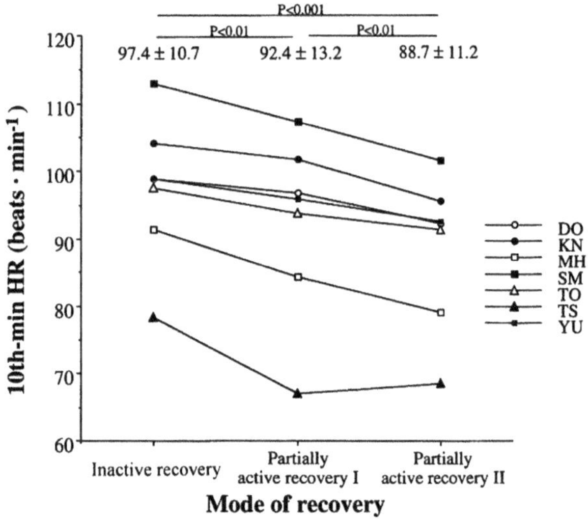

Fig. 3.2 Individual HR (heart rate) and mean ± SD values for seven subjects at the tenth minute of recovery. Adapted from Takahashi et al. 2000a, reproduced by permission

exercise, heart rates decreased rapidly during the first 2 min and decreased gradually thereafter, regardless of recovery mode. The mean heart rates at the sixth minute were 104.1 ± 10.3 beats/min during inactive recovery and 104.6 ± 12.7 beats/min during active recovery, between which the difference was insignificant. After 6 min of recovery exercise, however, resting heart rate decreased promptly and then remained below the corresponding heart rate for inactive recovery. In all six participants, the resting heart rate from the seventh through ninth minutes (the last 3 min) of recovery was significantly lower after cool-down exercise than after complete rest: 92.1 ± 7.4 vs. 100.8 ± 8.9 beats/min, $P < 0.05$, Table 3.1.

Figure 3.3b shows spectra of R-R interval variability during the last 3 min of recovery in the same participant as shown in Fig. 3.3a. The amplitude of the high-frequency component of R-R interval variability appeared to be greater at rest after cool-down exercise than at rest without cool-down exercise. Evidently, in all six subjects, the group mean value for high-frequency amplitude after cool-down exercise was slightly, but significantly ($P < 0.05$), greater than that without cool-down exercise (Table 3.1). Furthermore, the high-frequency amplitude was significantly greater after cool-down exercise than during cool-down exercise (10.6 ± 5.3 versus 5.6 ± 4.7 ms, $P < 0.05$). Across inactive recovery, however, the high-frequency amplitude for the seventh through ninth minutes did not significantly increase from that for the fifth through sixth minutes (9.0 ± 4.7 versus 7.4 ± 6.1 ms, respectively).

Figure 3.4 shows changes in the depth (tidal volume) and rate (respiratory frequency) of breathing during the last 3 min of recovery in the same participant as shown in Fig. 3.3. The mean values for tidal volume and respiratory frequency after cool-down exercise were similar to those without cool-down exercise. In all six participants, there were no significant differences in either tidal volume or respiratory frequency between the different modes of recovery (Table 3.1).

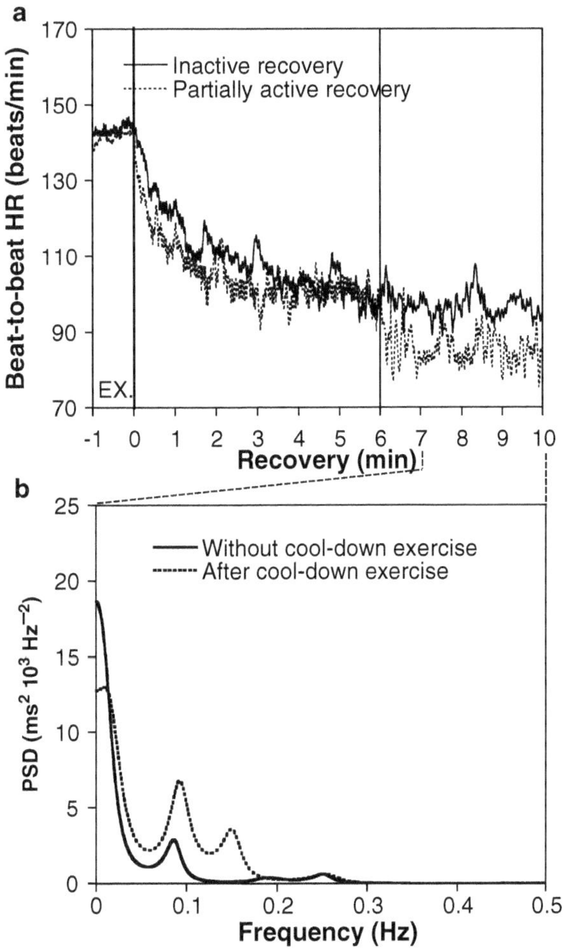

Fig. 3.3 (**a**) Changes in heart rate (HR) during 10 min of recovery from 5-min bout of exercise (Ex.) at 70% of peak oxygen uptake in a representative subject. The two modes of recovery were inactive recovery (complete rest) and partially active recovery (rest following 6 min of cycling at relative intensity of 20% of peak oxygen uptake). Vertical thick line (at $t = 0$) represents the termination of exercise and vertical thin line the termination of cycling. (**b**) Power spectral density (PSD) of R-R interval variability during quietly resting period of the last 3 min (from minutes 7 to 10) during recovery in the same subject. Adapted from Takahashi et al. 2002, reproduced by permission

Table 3.1 Amplitude of high-frequency component of heart rate variability and cardiorespiratory variables in resting states without and after cool-down exercise

		Without cool-down	After cool-down	*P* value
HR	beats/min	100.8 ± 8.9	92.1 ± 7.4	$P < 0.05$
R-R interval	ms	599 ± 51	656 ± 56	$P < 0.05$
HF amplitude	ms	9.0 ± 4.7	10.6 ± 5.3	$P < 0.05$
LF amplitude	ms	26.0 ± 20.3	31.1 ± 18.2	NS
LF/HF		2.8 ± 1.1	3.0 ± 0.7	NS
V_T	liters/breath	0.53 ± 0.09	0.56 ± 0.10	NS
f	breaths/min	19.2 ± 2.4	17.4 ± 1.8	NS
	hertz	0.32 ± 0.04	0.29 ± 0.03	NS

Values are presented as mean ± SD for six subjects
HR heart rate, *HF* high frequency, *LF* low frequency, *LF/HF* LF to HF ratio, V_T tidal volume, f respiratory frequency, *ms* millisecond
Adapted from Takahashi et al. 2002, reproduced by permission

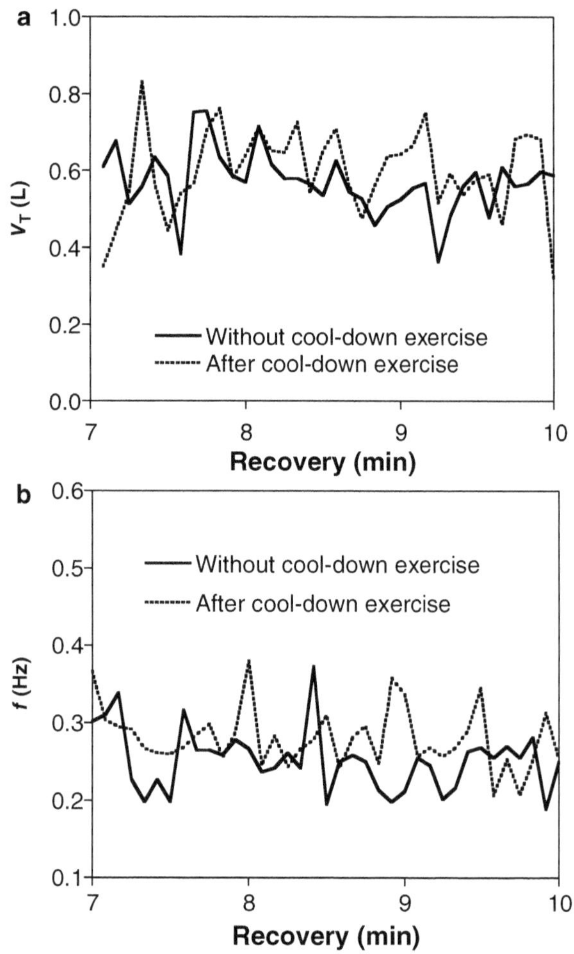

Fig. 3.4 (**a**) Changes in tidal volume (V_T, in liters per breath) and (**b**) respiratory frequency (*f*, in hertz) during the last 3 min of recovery in the same subject as show in Fig. 3.3. Adapted from Takahashi et al. 2002, reproduced by permission

3.4 Discussion

3.4.1 Experiment 1

3.4.1.1 Importance of Cool-Down Exercise

After heavy exercise, heart rate returning toward the resting level was enhanced by cool-down exercise at a light intensity during recovery. These results are entirely consistent with those of our previous study. Furthermore, the magnitude of reduction in heart rate after a cool-down exercise appears to be dependent on the duration of the cool-down exercise; that is, the longer the cool-down exercise, the lower the heart rate.

Cool-down exercise is widely believed to provide a latent protection against sudden cardiac death and to remove fatigue-related substances, such as lactic acid, from the working muscles and the circulation. Some study has suggested that the best strategy during recovery from exercise is to diminish the work load gradually to reduce the risk of cardiac events. Postexercise cardiovascular responses are mediated by the sustained activity of sympathetic efferent nerves to the heart and peripheral vessels. The sympathetic nerves originate from muscle metaboreceptor reflexes, arterial baroreceptor reflexes, or both. During such periods of high sympathetic activity, prolonged elevation of postexercise heart rate above the basal resting level may induce fatal cardiac arrhythmia and sudden cardiac death in vulnerable individuals with silent heart disease. Thus, understanding how cardiac functions, particularly the rate of heart beats, and how it is affected by a difference in the duration of the cool-down exercise is of crucial importance in preventing the risk of cardiac events during intervals between physical activities or competitive sports as well as during recovery from exercise. To our knowledge, our previous research was probably the first to demonstrate the facilitative effects of cool-down exercise on heart rate returning toward the pre-exercise resting level. This is also the first study to show that the degree of reduction in recovery heart rate following cool-down exercise depends on the length of the cool-down exercise. Although cool-down exercise is beneficial to return postexercise heart rate to the pre-exercise resting level, the mechanisms that facilitate recovery of postexercise heart rate at rest after active-recovery exercise are still unclear.

3.4.1.2 Dependence of Reduced Heart Rate on Duration of Active Recovery

Taken from the results of another study of ours, Fig. 3.5 shows the changes in heart rate and cardiac output during active recovery with different lengths of cool-down exercise (1-, 2-, and 3-min unloaded pedaling) and during quietly resting recovery from 5-min cycle exercise at a moderate intensity. The time courses of heart rate during a period of 3 min were almost identical, regardless of recovery mode (*upper panel* in Fig. 3.5). After cessation of unloaded pedaling of both 2 and 3 min, however, both heart rates at subsequent rest further decreased and reached their respective constant levels. The mean (±SD) values for resting heart rate at the fifth minute for inactive recovery (non-pedaling) and active recovery (1-, 2-, and 3-min pedaling) were 96.2 ± 5.2, 95.5 ± 6.3, 89.7 ± 6.4, and 89.0 ± 5.4 beats/min, respectively. The difference in heart rate was significant ($P < 0.05$) between non-pedaling and 3-min pedaling recovery. Especially, the order of decrease in heart rate from the third through the fifth minute was in the following: 3-min, 2-min, and 1-min pedaling.

The time courses of cardiac output under the four recovery conditions also varied with different periods of pedaling (*lower panel* in Fig. 3.5). The mean cardiac outputs at the third minute were significantly ($P < 0.05$) greater in pedaling participants than in non-pedaling participants. The stroke volumes during the three pedaling

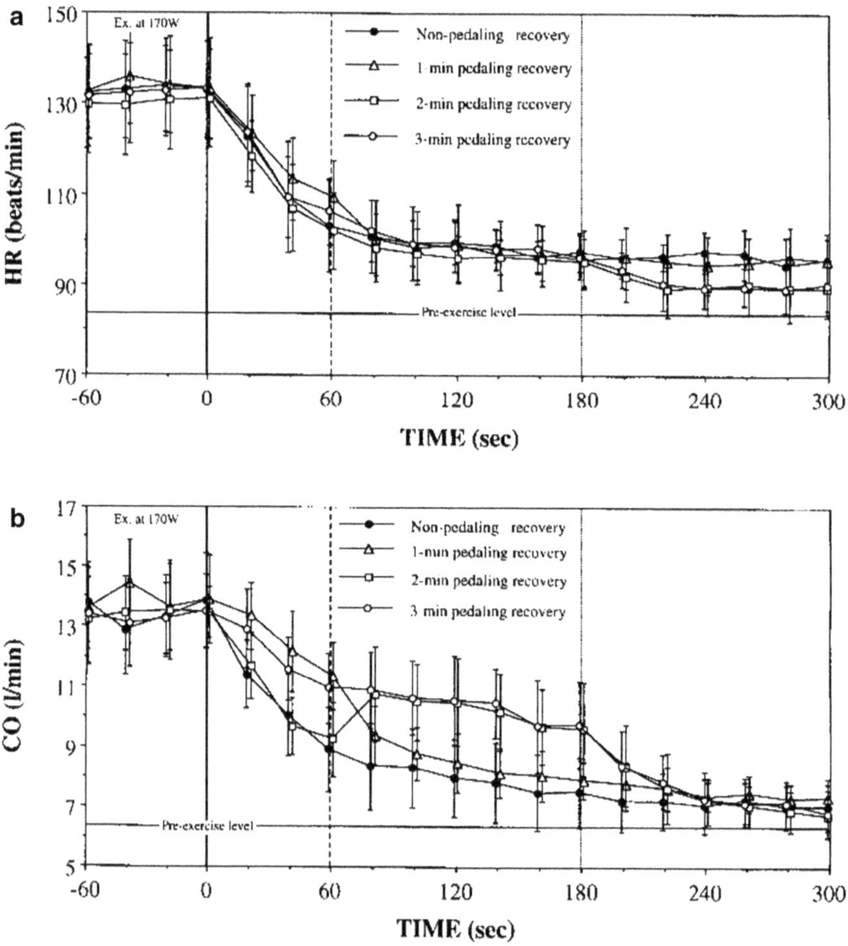

Fig. 3.5 Heart rate (HR, **a**) and cardiac output (CO by impedance cardiography, **b**) during recovery with non-pedaling and 1, 2, and 3 min of unloaded pedaling from upright cycle exercise (Ex.) of 5 min at 170 watts after seated rest of 5 min on an ergometer. Each symbol is represented as mean ± SD obtained from six healthy untrained young male subjects. The four different modes of recovery consisted of the following conditions: (1) complete sitting rest for 5 min (non-pedaling recovery); (2) unloaded pedaling for 1 min followed by 4 min of rest (1-min pedaling recovery); (3) rest for 1 min and then unloaded pedaling for 2 min followed by 2 min of rest (2-min pedaling recovery); and (4) unloaded pedaling for 3 min followed by 2 min of rest (3-min pedaling recovery). Adapted from Takahashi et al. 1998, reproduced by permission

periods were maintained at levels similar to those during the preceding moderate exercise. That is why cardiac output at the third minute is greater for a 3-min pedaling than for non-pedaling because of almost no difference in heart rate between the two. Thus, rhythmical muscular contractions of the lower extremities during recovery from exercise play important roles in promoting the venous return of blood to the heart in the upright position. They also prevent the venous blood from pooling, which is induced by the effect of gravity, in the legs with postexercise vasodilation.

The current study has observed the changes in heart rate during recovery with cool-down exercise after heavy exercise. We should also consider the physiological significance of cool-down exercise on the heart during the periods of cool-down exercise and subsequent rest. Some biochemical studies have suggested that the antiarrhythmic effect of mutual antagonism between plasma potassium and norepinephrine on the heart is maintained during an early phase of active recovery with continuous muscular contractions. The risk of cardiac arrhythmia during the subsequent slower phase of heart rate recovery may also be reduced after active recovery in comparison to inactive recovery. The decreased heart rate at rest after cool-down exercise would be elicited by an increase in parasympathetic nerve activity, a decrease in sympathetic nerve activity, or both.

Heart rate during recovery, specifically for a person performing moderate or heavy exercise, remains elevated above the resting base-line level for a relatively long period, when compared to ventilation. The present study only provides evidence that cool-down exercise has the effect of promoting heart rate recovery, but the mechanism remains unclear. Although the results of this study do not clarify the mechanism, they reveal that a protocol of cool-down exercise during recovery facilitates heart rate recovery from heavy exercise and that the promoted recovery of heart rate appears to be dependent to some extent on the duration of the cool-down exercise.

3.4.2 Experiment 2

3.4.2.1 Cardiac Vagal Activity After Cool-Down Exercise

Numerous studies on the relation of the autonomic nervous system to the heart give us some hints that alterations in cardiac efferent vagal activity play a primary role in the chronotropic responses (heart contraction rates) both at rest and during light-intensity exercise. Accordingly, we hypothesized that an increase in the tonic level of cardiac vagal reactivation during recovery from exercise should be responsible for a decrease in the postexercise level of heart rate after cool-down exercise of light intensity. In accordance with the results of our previous studies, we recognized in this study that heart rate during recovery from high-intensity exercise was significantly reduced after the participants performed cool-down exercise. This study also demonstrated that the high-frequency amplitude of R-R interval variability was slightly, but significantly, greater at rest after cool-down exercise than either during cool-down exercise or during inactive recovery without cool-down exercise. These results suggest that the high-frequency amplitude is closely related to the decline in the mean levels of heart rate toward the pre-exercise resting levels during recovery from exercise.

Our previous study in Chap. 2 has investigated how the recovery of the cardiovascular system differs between upright and supine positions from the same upright heavy exercise on a cycle ergometer. Each of the participants had a longer R-R interval and a greater R-R interval variability during recovery in the supine than in the upright position; that is, the longer the R-R interval (or the lower the heart rate), the greater the amplitude of high-frequency component of heart rate variability. These results have suggested that increased cardiac vagal activity is responsible for heart rate returning toward its basal resting level during recovery from exercise. Consistent with our hypothesis, the present study has also demonstrated that the high-frequency amplitude of R-R interval variability during recovery is significantly increased after the performance of cool-down exercise compared with complete rest without cool-down exercise. What is more important is that the vagal outflow to the heart, particularly during the slower phase of recovery, is enhanced as the aftereffect of a cool-down exercise. Our interpretation of one of the mechanisms for the enhancement of heart rate recovery is supported by the finding that cool-down exercise causes the tonic level of cardiac vagal efferent to increase, as shown in the increased high-frequency amplitude after cessation of cool-down exercise; that is, the contribution of postexercise cardiac vagal reactivation to the heart is enhanced by cool-down exercise. Thus, the significant decrease in heart rate following cool-down exercise can be partly explained by this tonic vagal mechanism.

3.4.2.2 Sympathetic Outflow to the Heart

A question on which we have to focus here is whether withdrawal of cardiac sympathetic nerve efferent contributes to the decrease in heart rate after the end of cool-down exercise. Other studies of animal exercise have demonstrated that sympathetic control of heart rate is secondary to vagal control both at rest and during exercise. In fact, the prompt reductions in heart rate that occur immediately after the end of cool-down exercise in this study are consistent with the results of rapid changes in heart rate responses to vagal nerve stimulation rather than cardiac sympathetic retreat in those studies. Someone else's research has demonstrated that the sympathetic nervous system that was triggered by the impulse traffic of group IV nerve fibers from the working muscles could control the state of elevated mean blood pressure via a muscular chemoreflex; thus, arterial blood pressure during the slow phase of recovery appears to be controlled through the sympathetic vasoconstriction by the chemoreflex in postexercise resting states. By contrast, the contribution of cardiac sympathetic withdrawal may be trivial to the heart.

3.4.2.3 Mechanisms of Long-Term Elevation of Heart Rate

As for the long-term elevation of heart rate during the slower phase of inactive recovery, the underlying mechanisms responsible for this phenomenon can be partly explained by neurohumoral effects. As revealed in the existing literature, several sympathetic co-transmitters, such as neuropeptides, released from sympathetic efferent nerves to the heart during the antecedent exercise attenuate vagal action at the heart during recovery. This neurohumoral influence is proved by the observations of other researchers that in the course of time a decrease in neuropeptide Y level from the heart parallels the inhibition of vagal control of heart rate responses up to tens of minutes after cardiac sympathetic nerve stimulation; however, the overflow of norepinephrine returns to its prestimulation level in 3 min.

Some investigators have reported that vagal activation effectively prevents ventricular fibrillation caused by acute myocardial ischemia, for example. As shown in this study, cardiac vagal efferent is probably being recuperated from suppression of parasympathetic neuronal input with a protocol of cool-down exercise. The cool-down exercise is therefore likely to play an important role in reducing lethal arrhythmia in vulnerable individuals with silent heart disease; see the Discussion section in Chap. 2.

3.4.2.4 Limitations

Several studies have demonstrated that increases in heart rate in response to exercise are attributable to withdrawal of cardiac parasympathetic nerve efferents, activation of cardiac sympathetic nerve efferents, or both. Accordingly, the balance between the two divisions of the autonomic nervous system seems likely to be different for the heart, particularly under the different conditions of active and inactive recovery. It is, however, difficult to describe the mechanisms that lead to similar heart rate deceleration during the early phase of recovery between the two different modes. Unfortunately, we have not yet built any sufficient database regarding the balance or interaction of the autonomic nervous system during recovery from exercise.

The mean values for tidal volume and respiratory frequency were not statistically different between the two resting states of recovery with and without cool-down exercise. According to other studies, usually, both the depth and rate of breathing must be consciously controlled to certain values to estimate cardiac vagal nerve activity from the amplitude of the high-frequency component of heart rate variability. However, there have been shown to be both interindividual and intraindividual differences in respiratory rhythm and pattern to maintain respiratory homeostasis from the disturbance of the biological system caused by intensive exercise. Thus, control of breathing such as a constant respiratory frequency may result in non-physiological states during the recovery period, particularly after a heavy exercise.

Breathing depth and rate usually show a reciprocal relationship to maintain partial pressure of arterial carbon dioxide at a physiological level (e.g., 40 mmHg) for any given constant ventilation. If participants are not allowed self-controlled respiratory rhythm, they should regulate end-tidal partial pressure of carbon dioxide with an increase or a decrease in tidal volume; however, the depth of breathing rather affects the parasympathetic tone. Although the participants made no attempt to control their breathing rate in this study, they happened to breathe at a rate of 0.15 Hz or higher, and these values were included in the evaluation component of the power spectral analysis. There is still some uncertainty that the close relationship between the high-frequency amplitude and the mean heart rate should be attributed to specifically vagal-mediated regulation as a whole.

3.5 Summary

3.5.1 Experiment 1

To clarify the effect of differences in the duration of cool-down exercise on recovery heart rate after exercise, seven healthy young men underwent cycle exercise at 70% of peak oxygen uptake for 5 min in an upright seated position. Heart rate in the phase of recovery was evaluated under three different conditions: (1) complete rest for 10 min (inactive recovery); (2) cycling at 20% of peak oxygen uptake for 3 min followed by 7-min rest (partially active recovery I); and (3) cycling at 20% of peak oxygen uptake for 6 min followed by 4-min rest (partially active recovery II). Heart rate during the first 2 min of inactive recovery decreased rapidly, then gradually decreased, and leveled off to 97.4 ± 10.7 beats/min, which was significantly above its pre-exercise resting level. Similar decreases in heart rate were observed during the first 2 min of the two partially active recovery conditions at light-intensity cycling of 3 and 6 min. After cessation of active-recovery exercise, each resting heart rate evidently decreased to 92.4 ± 13.2 beats/min for partially active recovery I and 88.7 ± 11.2 beats/min for partially active recovery II at the tenth minute; these values were significantly different from each other and were both significantly lower than that for inactive recovery. From these results we can conclude that the restoration of slowly decreasing heart rate during recovery from heavy exercise is facilitated by cool-down exercise and this facilitated reduction in heart rate is dependent on the duration of the cool-down exercise.

3.5.2 Experiment 2

The recovery of postexercise heart rate is enhanced with cool-down exercise; however, the mechanisms involved in the aftereffect of active-recovery exercise on the prompt reduction in heart rate have not been fully understood. To determine whether an increase in the activity of cardiac vagal efferent nerves is associated with decreased heart rate after cool-down exercise, we assessed the amplitude of high-frequency component of R-R interval variability between 0.15 and 0.40 Hz, as an index of cardiac vagal tone. Six healthy young male participants who performed cycle exercise at 70% of peak oxygen uptake for 5 min and then recovered either during 4-min sitting rest after 6 min of cool-down cycling at 20% of peak oxygen uptake or during complete sitting rest throughout 10 min without cool-down exercise. After heavy exercise, heart rate decreased rapidly followed by its slow decline during the first 6 min of recovery, regardless of whether the participants performed cool-down exercise or not. However, the resting heart rate after cool-down exercise was significantly lower than that without cool-down exercise (92.1 ± 7.4 versus 100.8 ± 8.9 beats/min). At the time, the high-frequency amplitude was slightly, but significantly greater with than without cool-down exercise (10.6 ± 5.3 versus 9.0 ± 4.7 ms, by Wilcoxon signed-ranks test). Thus, the largely increased high-frequency amplitude corresponded to the decreased heart rate at rest after cool-down exercise, as compared with complete rest without cool-down exercise. Therefore, we conclude that after heavy exercise, the heart rate descending toward the pre-exercise resting level after cool-down exercise is associated with an increase in cardiac efferent vagal tone.

Further Reading

Takahashi T, Okada A, Hayano J, Tamura T, Miyamoto Y. Influence of duration of cool-down exercise on recovery of heart rate in humans. Ther Res. 2000a;21:170–5.

Takahashi T, Okada A, Hayano J, Tamura T. Influence of cool-down exercise on autonomic control of heart rate during recovery from dynamic exercise. Front Med Biol Eng. 2002;11:249–59.

Takahashi T, Niizeki K, Miyamoto Y. The restorative effect of light exercise duration on heart rate during recovery from exercise in man. In: Nose H, Nadel ER, Morimoto T, editors. The 1997 Nagano symposium on sports sciences. Cooper, Carmel: I. L; 1998. p. 59–63.

Arai Y, Saul JP, Albrecht P, Hartley LH, Lilly LS, Cohen RJ, Colucci WS. Modulation of cardiac autonomic activity during and immediately after exercise. Am J Phys. 1989;256:H132–41.

Bannister SR, Mathias CJ. Autonomic failure: a textbook of clinical disorders of the autonomic nervous system. New York: Oxford University Press; 1992.

Brown TE, Beightol LA, Koh J, Eckberge DL. Important influence of respiration on human R-R interval power spectra is largely ignored. J Appl Physiol. 1993;75:2310–7.

Caplin JL, Banim SO. Chest pain and electrocardiographic ST-segment elevation occurring in the recovery phase after exercise in a patient with normal coronary arteries. Clin Cardiol. 1985;8:228–9.

Dixon EM, Kamath MV, McCartney N, Fallen EL. Neural regulation of heart rate variability in endurance athletes and sedentary controls. Cardiovasc Res. 1992;26:713–9.

Esler M, Jennings G, Korner P, Willett I, Hasking F, Anderson W, Lambert G. Assessment of human sympathetic nervous system activity from measurements of norepinephrine turnover. Hypertension. 1988;11:3–20.

Furlan R, Piazza S, Dell'Orto S, Gentile E, Cerutti S, Pagani M, Malliani A. Early and late effects of exercise and athletic training on neural mechanisms controlling heart rate. Cardiovasc Res. 1993;27:482–8.

Goldberg DI, Shephard RJ. Stroke volume during recovery from upright bicycle exercise. J Appl Physiol. 1980;48:833–7.

Guyton AC, Douglas BH, Langston JB, Richardson TQ. Instantaneous increase in mean circulatory pressure and cardiac output at onset of muscular activity. Circ Res. 1962;11:431–41.

Hayano J, Mukai S, Sakakibara M, Okada A, Takata K, Fujinami T. Effects of respiratory interval on vagal modulation of heart rate. Am J Phys. 1994;267:H33–40.

Kamath MK, Fallen EL. Power spectral analysis of heart rate variability: a noninvasive signature of cardiac autonomic function. Crit Rev Biomed Eng. 1993;21:245–311.

Kantona PG, Jih F. Respiratory sinus arrhythmia: noninvasive measure of parasympathetic cardiac control. J Appl Physiol. 1975;39:801–5.
Komaru T, Ashikawa K, Kanatsuka H, Sekiguchi N, Suzuki T, Takishima T. Neuropeptide Y modulates vasoconstriction in coronary microvessels in the beating canine heart. Circ Res. 1990;67:1142–51.
Lahiri A, Balasubramanian V, Millar Craig MW, Crawley J, Raftery EB. Exercise-induced ST segment elevation electrocardiographic, angiographic, and scintigraphic evaluation. Br Heart J. 1980a;43:582–8.
Lahiri A, Subramanian B, Millar-Craig FM, Crawley MJ, Raftery EB. Exercise-induced S-T segment elevation in variant angina. Am J Cardiol. 1980b;45:887–94.
Lauf U, Urhausen A, Werner N, Scharhag J, Heitz A, Kissner G, Böhm M, Kindermann W, Nickenig G. Running exercise of different duration and intensity: effect on endothelial progenitor cells in healthy subjects. Eur J Cardiovasc Prev Rehabil. 2005;12:407–14.
Laughlin MH. Skeletal muscle blood flow capacity: role of muscle pump in exercise hyperemia. Am J Phys. 1987;253:H9933–1004.
Linnarsson D. Dynamics of pulmonary gas exchange and heart rate changes at start and end of exercise. Acta Physiol Scand Suppl. 1974;415:1–68.
Malik M. Clinical guide to cardiac autonomic tests. London: Kluwer Academic Publishers; 1998.
McCaffrey FM, Braden DS, Strong WB. Sudden cardiac death in young athletes: a review. Am J Dis Child. 1991;145:177–83.
Northcote RJ, Fannigan C, Ballantyne D. Sudden death and vigorous exercise: a study of 60 deaths associated with squash. Br Heart J. 1986;55:198–203.
O'Leary D. Autonomic mechanisms of muscle metaboreflex control of heart rate. J Appl Physiol. 1993;74:1748–54.
Opie LH. Sudden death and sport. Lancet. 1975;305:263–6.
Opie LH. Heart physiology: from cell to circulation. Philadelphia: Lippincott Williams & Wilkins; 2004.
Perini R, Orizo C, Baselli G, Cerutti S, Veicsteinas A. The influence of exercise intensity on the power spectrum of heart rate variability. Eur J Appl Physiol. 1990;61:143–8.
Pomeranz M, Macaulay RJB, Caudill MA, Kutz I, Adam D, Gordon D, Kilborn KM, Barger AC, Shannon DC, Choen RJ, Benson M. Assessment of autonomic function in humans by heart rate spectral analysis. Am J Phys. 1985;248:H151–3.
Potter EK. Effects of neuropeptides on the vagal neuroeffector junction. In: Levy MN, Schwartz PJ, editors. Vagal control of the heart: experimental basis and clinical implications. New York: Futura; 1994. p. 289–303.
Robinson BF, Epstein SE, Beiser GD, Braunwald E. Control of heart rate by the autonomic nervous system. Studies in man on the interrelation between baroreceptors mechanisms and exercise. Circ Res. 1966;19:400–11.
Rowell LB. Human cardiovascular control. New York: Oxford University Press; 1993.
Sawada Y, Ohtomo N, Tanaka Y, Tanaka G, Yamakoshi K, Terachi S, Shimamoto K, Nakagawa M, Satoh S, Kuroda S, Imurai O. New technique for time series analysis combining the maximum entropy method and non-linear least squares method: its value in heart rate variability analysis. Med Biol Eng Comput. 1997;35:318–22.
Stramba-Badiale M, Vanoli E, De Ferrari GM, Cerati D, Foreman ED, Schqartz PJ. Sympathetic-parasympathetic interaction and accentuated antagonism in conscious dogs. Am J Phys. 1991;260:H335–40.
Takahashi T, Miyamoto Y. Influence of light physical activity on cardiac responses during recovery from exercise in humans. Eur J Appl Physiol. 1998;77:305–11.
Takahashi T, Niizeki K, Miyamoto Y. Respiratory responses to passive and active recovery from exercise. Jpn J Physiol. 1997;47:59–65.
Takahashi T, Okada A, Saitoh T, Hayano J, Miyamoto Y. Difference in human cardiovascular response between upright and supine recovery from upright cycle exercise. Eur J Appl Physiol. 2000b;81:233–9.

Takahashi T, Yanagida H, Niizeki K, Wang S, Matsuo T, Watanabe S, Okada A, Kuwayama T, Natsui J. Economical estimates of oxygen uptake as a function of gait parameters for an ambulatory monitoring system. Ther Res. 2009;30:219–25.

Task Force of the European Society of Cardiology and the North American Society of Pacing and Electrophysiology. Heart rate variability. Standards of measurement, physiological interpretation, and clinical use. Circulation. 1996;93:1043–65.

Vanoli E, De Ferrari GM, Stramba-Badiale M, Hull SS, Foreman RD Jr, Schwartz PJ. Vagal stimulation and prevention of sudden cardiac death in conscious dogs with a healed myocardial infarction. Circ Res. 1991;68:1471–81.

Waner HR, Cox A. A mathematical model of heart rate control by sympathetic and vagus efferent information. J Appl Physiol. 1962;17:349–55.

Waner MR, Senanayake PDS, Ferrario CM, Levy MN. Sympathetic stimulation-evoked overflow or norepinephrine and neuropeptide Y from the heart. Circ Res. 1991;69:455–65.

Wang JS, Jen CJ, Kung H, Lin LJ, Hsiue TR, Chen H. Different effects of strenuous exercise and moderate exercise on platelet function in men. Circulation. 1994;90:2877–85.

Chapter 4
Cardiopulmonary Mechanoreflexes

4.1 Introduction

Some studies have demonstrated that an initial complex response of the heart rate, i.e., a sharp increase and a subsequent level off or slight decrease, at the onset of voluntary exercise, is induced by dynamic muscle contractions as standing, handgrip exercise, and leg exercise, but not induced by passive tilt standing. The initial fluctuation in heart rate during the early phase of physical activity has been reported to be attributable to the initial withdrawal of vagal efferent nerve activity and transient retrieval of vagal efferent nerve input in sequence. These studies have also demonstrated a complex response of the heart rate or an overshoot at the onset of upright exercise at almost only light to mild workloads; however, if a person exercises a heavy workload, the complex change in heart rate at the beginning hardly appears but rather increases exponentially. So far, little is known about the mechanisms whereby the heart rate is sustained below its initial peak or suppressed from a gradual increase. Such a fluctuation in heart rate is a very interesting phenomenon, highlighting the significance of circulatory function, because light physical activity is often inherent on a daily basis.

This chapter describes the biphasic change in heart rate that takes place in upright posture at the onset of cycle exercise at light intensity, but that does not occur in the supine posture. Differences in heart rate, stroke volume, cardiac output, and vagal nerve activity in response to dynamic exercise at light intensity between an upright and a supine position are also discussed. In addition to these biological variables, mean arterial pressure and heart rate variability (or R–R interval variability) are noninvasively measured beat-by-beat. R–R interval variability in a time domain is translated into a high-frequency component in a frequency domain in power spectral analysis by use of a maximal entropy method. The amplitude of a high-frequency component can be used for assessment of vagal nerve efferent to the heart, both at rest and during exercise.

T. Takahashi, *The Cardiovascular System During Exercise and Recovery*, https://doi.org/10.1007/978-981-99-8486-2_4

The blood inflow to and outflow from the heart must be essentially equal. In this regard, both stroke volume and central blood volume or central venous pressure will be proportional to the amount of venous return with alterations in body position and in muscle mass of dynamic contractions. Indeed, from rest to exercise in an upright position, the mechanical effect of the muscle pumping action effectively promotes venous blood in the centripetal direction from the lower extremities, hence this effect increases the central venous volume and pressure. Thus, as a result of increased cardiac filling, stroke volume is augmented via the Frank-Starling mechanism during the transition from rest to dynamic exercise, particularly in the upright position.

The objective of this study is also to evaluate the hypothesis that the vagal efferent nerve tone to the heart during upright exercise is mediated by the activity of cardiopulmonary mechanosensory receptors in the atria, ventricles, pulmonary artery, and pulmonary veins. Notably, the neural afferents from cardiopulmonary mechanoreceptors to the central nervous system would be induced by an increase in central blood volume, which results from venous return promoted by the skeletal muscle pump. When subjects perform dynamic exercise at light intensity, a gradually reduced heart rate after its initial rapid increase is maintained just above the pre-exercise resting level during exercise in an upright sitting position. This reduction in heart rate will be explained by the autonomic mechanisms that cardiac vagal control of heart rate is evoked by the symphathoinhibitory reflex of cardiopulmonary mechanoreceptors. Conversely, such a decline in heart rate would not be expected during exercise in a supine position, since almost no skeletal muscle pump can operate effectively with already elevated central venous volume.

4.2 Materials and Methods

4.2.1 Human Subjects

Eight young men, healthy untrained subjects, participated in this study. The physical characteristics of the subjects in 22.2 ± 1.0 (mean ± SD) years of age were a mean height of 1.73 ± 0.07 m, a mean body mass of 64.0 ± 7.0 kg, and a mean peak oxygen uptake of 2.77 ± 0.14 L/min. They were fully informed about the procedures, risks, and benefits of the study and provided written informed consent before the experiments. All the subjects were examined by a cardiologist, so they had no signs of cardiovascular disease. The experimental protocol was approved by the university institutional review board. The research and experiment was also carried out in compliance with the principles outlined in the Declaration of Helsinki regarding experimentation with human participants.

Peak oxygen uptake was determined for each subject, who performed an exercise test at an incremental work rate of 15 W every minute on an electromagnetic-braked cycle ergometer (Corival 300; Lode, the Netherlands) until he was exhausted.

The peak oxygen uptake was defined as the highest oxygen uptake obtained as a mean value over a 10-s period of breath-by-breath measurements (see Chap. 1 for details).

4.2.2 Measurements of Cardiorespiratory Variables

The methods for measuring cardiovascular and respiratory variables are described in detail in the previous chapters; they are briefly outlined again here.

Tidal volume (V_T, in liters per minute), respiratory frequency (f, in breaths per minute), and oxygen uptake (VO_2, in liters per minute) were recorded on a breath-by-breath basis with a computerized measurement system. Ventilatory airflow was monitored with the use of a hot-wire-type pneumotachograph (RF-2; Minato, Osaka, Japan). The composition of expired gas was continuously analyzed with a medical mass spectrometer (WSMR-1400; Westron, Chiba, Japan), which was calibrated with a standard reference gas mixture before each measurement.

Heart rate (*HR*, in beats per minute), stroke volume (*SV*, in milliliters per beat), and cardiac output (*CO*, in liters per minute) were recorded with a computerized measurement system. R–R intervals, the time interval between two successive R peaks (which reflect depolarization of the left ventricle), were monitored by electrocardiography at the chest wall with bipolar standard leads. Heart rate was calculated as the quotient of 60 divided by the R–R interval. Stroke volume was determined by impedance cardiography (ICG). The transthoracic impedance was measured with a standard constant-current-type ICG unit (RGA-5; Nihon Koden, Tokyo) with four silver tape electrodes; one pair of them were placed around the neck and the other pair around the torso. Stroke volume was calculated from the Kubicek equation. Artifacts on impedance signals, due to the unfavorable influence of respiration and body motion, were removed arithmetically by averaging ICG waveforms over 10 cardiac cycles. The validity of stroke volume that is derived by the monitoring system of impedance cardiogram has been confirmed under various experimental conditions, including rest, exercise, and postural change. Although the absolute values of impedance-derived stroke volume remain to be elucidated, there has been a general agreement that ICG is suitable for estimating relative changes in stroke volume during exercise and during postural change. All data for respiratory and cardiac variables were stored in a memory device of a personal computer for subsequent analysis.

Systolic arterial blood pressure and diastolic blood pressure were measured noninvasively with the use of a pneumatic finger cuff (2300 Finapres; Omeda, Tokyo). Mean blood pressure (*MBP*, in millimeters of mercury, mmHg) was calculated as diastolic pressure plus one-third of the pulse pressure. The validity of finger blood pressure recorded in this way has been demonstrated by comparison with intra-arterial pressure measured at the same time under various conditions. The finger cuff was held at heart level to avoid errors due to hydrostatic pressure throughout all the experiments. The mean blood pressure and R–R interval data were recorded sequentially on a personal computer via an RS-232C interface.

Total peripheral vascular resistance (*TPR*, in dyn·s^{-1} cm^{-5}) was calculated by dividing mean blood pressure by cardiac output and multiplying by 80, where 1 mmHg L^{-1} min = 80 dyn s^{-1} cm^{-5}. The constant 80 is a conversion factor to change units to standard resistance units.

4.2.3 Experimental Protocols

Each subject performed the same exercise twice in two different postures, upright (sitting) and supine (lying), as shown in Fig. 4.1. The subject rested for 6 min and then performed cycle exercise at 30 W for 6 min. The subject maintained either the upright position on the cycle ergometer or the supine position on a bed behind the ergometer throughout the experiments. The subject placed his feet on the footrest near the flywheel during the period of rest and then pedaled at a constant rate of 60 revolutions/min paced by a metronome during exercise. The pedaling rate was monitored with an inductive revolution meter fitted to the ergometer. Subjects were allowed to breathe spontaneously; that is, no attempt was made to control the pattern, depth, or frequency of breathing throughout the experiments.

The subjects had exercise tests at about 13:00 on separate days. The order of the four runs in total was randomly assigned for the upright and supine exercise tests. The subjects were scheduled to participate in the experiments once a week so that no improvement of their exercise capacity by exercise training would affect the circulatory or respiratory responses to the exercise tests. They were not allowed to consume any beverages containing caffeine or alcohol from 21:00 the night before the day of exercise tests. Vigorous exercise was also forbidden for 48 h before the day of testing. The subjects had eaten a light meal at least 3 h before the tests. All experiments were conducted at a room temperature of 21–22 °C.

4.2.4 Power Spectral Analysis

Before spectral analysis, R–R intervals sampled at a rate of 500 hertz (Hz) were interpolated by a linear function, thereby yielding an equidistantly sampled R–R interval function. The rearranged R–R interval function during the last 2 min both at

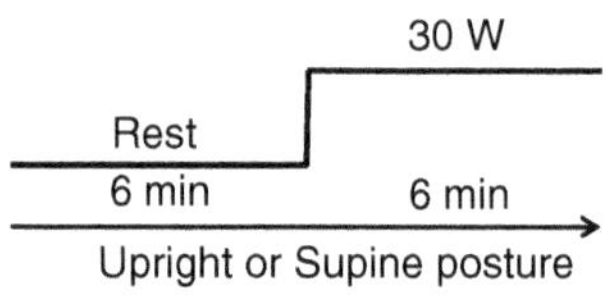

Fig. 4.1 Schematic representation of exercise protocols: rest for 6 min and then exercise at 30 W for 6 min in either upright posture or supine posture. Each posture was consistently maintained for 12 min

rest and during exercise was subjected to power spectral analysis by use of a maximal entropy method. Linear trends, if any, were removed from the R–R interval time series before spectral analysis. The data that were subjected to the above processing were analyzed with a commercially available computerized system (MemCal; Suwa-Torasuto, Tokyo). The validity of this spectral analysis system for human R–R interval data has been reported elsewhere in the literature.

The amplitude of a high-frequency (HF, 0.15–0.40 Hz) component of R–R interval variability was used to assess vagal nerve efferent activity to the heart. The high-frequency amplitude and R–R interval are both dependent on the firing rate of the efferent vagal nerves. An inverse relationship between high-frequency amplitude and heart rate then provides insight into alterations in vagal nerve tone to the sinoatrial node, which directly controls heart rate. Some studies have suggested that estimating the activity of cardiac sympathetic nerve efferents is relatively difficult from low-frequency components during exercise. Thus, variability other than the high-frequency component was not assessed in spectral analysis. In this study, the magnitude of the high-frequency component was expressed as mean amplitude, which was calculated as mean amplitude = $(2 \times \text{power of component})^{1/2}$.

4.2.5 *Data Analysis*

Cardiorespiratory data obtained from each of four runs were regularly interpolated on a time base of 5-s intervals with Lagrange polynomials. The two arranged data of variables from exercise tests performed in the same posture were then averaged to yield a single set of data for each subject. Group mean values for all variables were averaged from the eight individual sets of interpolated data. Arithmetic mean values for all variables were determined from the last minute at rest before exercise, from 20 to 30 s of exercise, and from the 3rd and 6th minutes of exercise. The amplitude of the high-frequency component of R–R interval variability was determined from the last 2 min of data recorded before and during exercise.

4.2.6 *Statistical Analysis*

Differences in the mean values of variables at rest and during exercise in each body position were evaluated by one-way analysis of variance with repeated measures and Scheffé's multiple comparison post hoc test. Differences in the changing patterns of variables between upright and supine postures were evaluated by two-way analysis of variance with repeated measures. Multiple comparison tests after the analysis of variance were performed to decide significant differences existing between any pair of all levels (time series points) by two-sided Student's paired t test. Differences in high-frequency amplitude, respiratory frequency, and tidal volume between rest and exercise and between upright and supine postures were also analyzed by use of the Friedman analysis of variance and Wilcoxon's signed-rank test adjusting the alpha level for multiple comparisons when necessary. Differences were considered significant for all statistical analyses at $P < 0.05$. Data are presented as mean ± standard error (SE).

4.3 Results

Figure 4.2 shows the temporal changes in oxygen uptake from rest to cycle exercise at a work rate of 30 W. The relative intensity of exercise was 20 ± 1% of peak oxygen uptake for the upright position and 21 ± 2% of peak oxygen uptake for the supine position. At the onset of exercise, oxygen uptakes increased immediately and remained at relatively stable levels in both positions. The exercising oxygen uptake was significantly ($P < 0.05$) greater than the resting oxygen uptake in each position. The time courses of oxygen uptake for the two positions were very similar, where no significant differences were detected between the two positions either at rest or during exercise (Table 4.1).

In the upright position, heart rate increased immediately after the start of exercise and then decreased gradually to just above the pre-exercise resting level; particularly, the entire response that was peculiar to upright exercise at light intensity was not exponential, for example (*upper panel* in Fig. 4.3). Such a reduction in heart rate following the initial rise was observed in every subject in the upright position, but not in the supine position. The mean peak of upright heart rate was 86.3 ± 1.4 beats/min at 25 ± 1 s. In contrast, at the onset of exercise in the supine position, heart rate increased rapidly to a relatively stable level without any apparent peak. The peak of heart rate in the upright position was significantly higher than the heart rate at the same time in the supine position (86.3 ± 1.4 vs. 78.6 ± 1.6 beats/min, $P < 0.05$, Table 4.1). However, the gradual decrease in heart rate after the peak in the upright

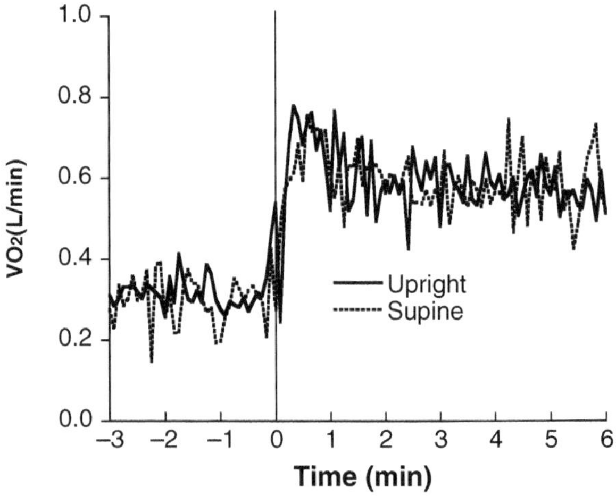

Fig. 4.2 Oxygen uptake (VO_2) at rest and during a 6-min bout of cycle exercise at 30 W in upright and supine positions. Vertical line (at $t = 0$) indicates the start of exercise. Data are group mean values obtained from eight subjects in both postures. Adapted with permission from Takahashi et al.

Table 4.1 Cardiovascular and respiratory responses to cycle exercise at 30 W in upright and supine positions

			Exercise		
	Posture	Rest	20–30 s	3 min	6 min
VO_2 (L/min)	Upright	0.31 ± 0.02	0.74 ± 0.05[a]	0.58 ± 0.03[a,b]	0.56 ± 0.02[a,b]
	Supine	0.30 ± 0.02	0.63 ± 0.06[a]	0.57 ± 0.04[a]	0.58 ± 0.04[a]
	Up vs. Su	n.s.	n.s.	n.s.	n.s.
HR (beats/min)	Upright	68.6 ± 1.2	86.3 ± 1.4[a]	72.1 ± 0.8[a,b]	71.9 ± 1.2[a,b]
	Supine	61.3 ± 1.4	78.7 ± 1.6[a]	77.4 ± 1.2[a]	77.2 ± 1.4[a]
	Up vs. Su	$P < 0.05$	$P < 0.05$	$P < 0.05$	$P < 0.05$
SV (mL)	Upright	73.5 ± 3.3	79.5 ± 2.3[a]	87.0 ± 2.5[a,b]	86.2 ± 2.8[a,b]
	Supine	107.5 ± 4.3	103.1 ± 6.2	100.3 ± 4.2	99.5 ± 3.1
	Up vs. Su	$P < 0.05$	$P < 0.05$	$P < 0.05$	$P < 0.05$
CO (L/min)	Upright	5.0 ± 0.2	6.7 ± 0.1[a]	6.3 ± 0.1[a,b]	6.2 ± 0.2[a,b]
	Supine	6.6 ± 0.3	8.1 ± 0.5[a]	7.8 ± 0.3[a]	7.7 ± 0.3[a]
	Up vs. Su	$P < 0.05$	$P < 0.05$	$P < 0.05$	$P < 0.05$
MBP (mmHg)	Upright	85.7 ± 2.1	86.8 ± 2.1	87.7 ± 1.6	85.1 ± 2.1
	Supine	85.1 ± 2.3	80.6 ± 1.3	89.8 ± 1.8[b]	89.2 ± 1.8[b]
	Up vs. Su	n.s.	n.s.	n.s.	n.s.
TPR (dyn s^{-1} cm^{-5})	Upright	1280 ± 211	962 ± 102[a]	1037 ± 66[a]	1013 ± 86[a]
	Supine	1036 ± 146	814 ± 151[a]	921 ± 91[a,b]	922 ± 90[a]
	Up vs. Su	$P < 0.05$	$P < 0.05$	$P < 0.05$	$P < 0.05$
HF (ms)	Upright	25.5 ± 5.1	–	–	42.2 ± 7.1[a]
	Supine	48.3 ± 6.1	–	–	23.9 ± 2.6[a]
	Up vs. Su	$P < 0.05$			$P < 0.05$
f (breaths/min)	Upright	14.5 ± 0.9	–	–	17.5 ± 1.6
	Supine	15.9 ± 1.3	–	–	16.1 ± 1.1
	Up vs. Su	n.s.			n.s.
V_T (L)	Upright	0.75 ± 0.10	–	–	0.84 ± 0.07
	Supine	0.61 ± 0.03	–	–	0.89 ± 0.08
	Up vs. Su	n.s.			n.s.

Values are means ± SE for eight subjects
Adapted with permission from Takahashi et al.
VO_2 oxygen uptake, *HR* heart rate, *SV* stroke volume, *CO* cardiac output, *MBP* mean blood pressure, *TPR* total peripheral vascular resistance, *HF* the amplitude of high-frequency component of R–R interval variability, *f* respiratory frequency, V_T tidal volume
[a] $P < 0.05$, significantly different from rest in the same position
[b] $P < 0.05$, significantly different from the mean value at 20–30 s

position was significantly lower than the corresponding heart rate in the supine position at the 6th minute (71.9 ± 1.2 vs. 77.2 ± 1.4 beats/min, $P < 0.05$).

As shown in the lower panel of Fig. 4.3, stroke volume at rest in the upright position was significantly lower by approximately 30% than in the supine position (73.5 ± 3.3 vs. 107.5 ± 4.3 mL, $P < 0.05$). Upright stroke volume increased significantly ($P < 0.05$) by approximately 20% from rest to exercise, whereas supine stroke volume decreased slightly, but insignificantly. The mean stroke volumes during

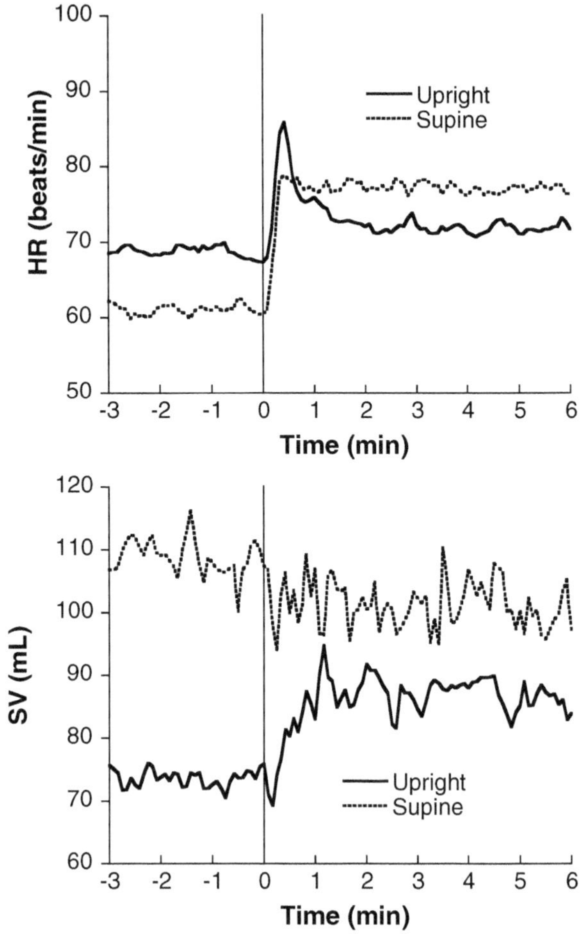

Fig. 4.3 Heart rate (HR) and stroke volume (SV) at rest and during 6 min of cycle exercise at 30 W in upright and supine positions. Vertical line (at *t* = 0) indicates the start of exercise. Data are group mean values obtained from eight subjects in both postures. Adapted with permission from Takahashi et al.

exercise were still significantly ($P < 0.05$) less in the upright than in the supine position at the 3rd and 6th minutes (Table 4.1).

The top panel in Fig. 4.4 shows the time courses of cardiac output from rest to exercise in the upright and supine positions. The cardiac output increased in a step-like manner after the start of exercise in both the positions. Upright cardiac output significantly ($P < 0.05$) increased from 5.0 ± 0.2 L/min at rest to 6.2 ± 0.2 L/min during exercise, and supine cardiac output significantly ($P < 0.05$) increased from 6.6 ± 0.3 L/min at rest to 7.7 ± 0.3 L/min during exercise. The differences in cardiac output between the two positions were significant (both $P < 0.05$) at rest and during exercise (Table 4.1). The increased amount of upright cardiac output from rest to exercise was similar to that of supine cardiac output by approximately 1.0 L/min; however, the relative increase in cardiac output was greater in the upright position by 24% than in the supine position by 17%.

Fig. 4.4 Cardiac output (CO), mean blood pressure (MBP), and total peripheral vascular resistance (TPR) at rest and during 6 min of cycle exercise at 30 W in upright and supine positions. Vertical line (at $t = 0$) indicates the start of exercise. Data are group mean values obtained from eight subjects in both postures. Adapted with permission from Takahashi et al.

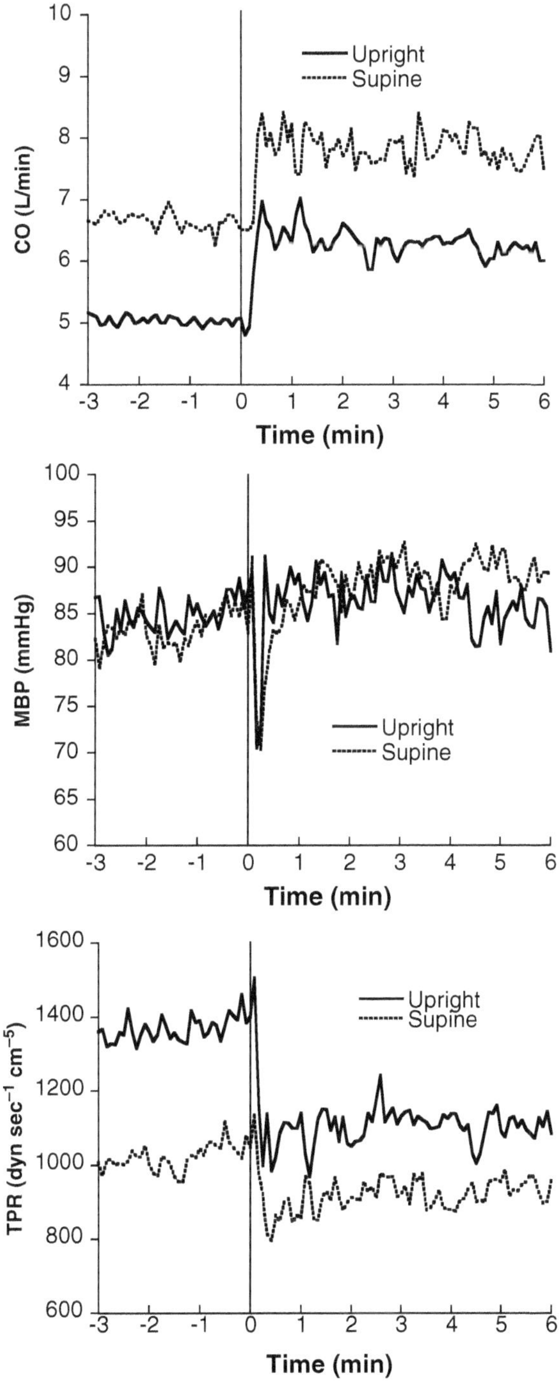

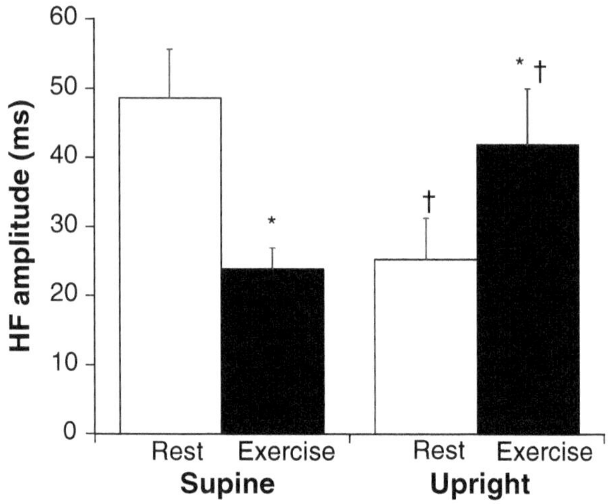

Fig. 4.5 Amplitude of high-frequency (HF, 0.15–0.40 Hz) component of R–R interval variability during the last 2-min periods of rest and exercise in upright and supine positions. Data are group mean values (mean ± SE) obtained from eight subjects in both postures. *$P < 0.05$, significantly different from rest in the same position; †$P < 0.05$, significantly different from supine values for rest and exercise. Adapted from Takahashi et al., reproduced by permission

Immediately after the start of exercise, mean blood pressures for both the positions decreased transiently and then returned to the pre-exercise levels (*middle panel* in Fig. 4.4); the minimum value was 70.5 ± 1.6 mmHg at 10 s in the upright and 70.4 ± 2.3 mmHg at 15 s in the supine position. The changes in mean blood pressure during the transition from rest to exercise were not significantly different between the two positions (Table 4.1).

The time courses of total peripheral vascular resistance from rest to exercise showed almost upside down against those of cardiac output in both the positions (*bottom panel* in Fig. 4.4). Total peripheral resistance during exercise in each body position was significantly ($P < 0.05$) lower than that at rest. The steady-state total peripheral resistances at rest and during exercise were significantly ($P < 0.05$) greater in the upright than in the supine position (Table 4.1).

The high-frequency amplitude of R–R interval variability in the supine position was significantly greater at rest than during exercise (48.3 ± 6.1 vs. 23.9 ± 2.6 ms, $P < 0.05$, Fig. 4.5). In contrast, the high-frequency amplitude in the upright position was significantly greater during exercise than at rest (42.2 ± 7.1 vs. 25.5 ± 5.1 ms, $P < 0.05$). The high-frequency amplitude at rest was significantly ($P < 0.05$) greater in the supine than in the upright position, whereas that during exercise was significantly ($P < 0.05$) greater in the upright position.

For respiratory variables, there were no significant differences in respiratory frequency or tidal volume between the upright and supine positions at rest and during exercise.

4.4 Discussion

4.4.1 Lower Heart Rate in Upright than Supine Position

As shown in Fig. 4.3, heart rate rapidly increased at the onset of exercise. This rapid increase in heart rate occurred in both the upright and supine positions. After the increase peaked, heart rate decreased gradually in the upright seated position and reached a steady level in 3 min during exercise. This steady-state value was slightly higher than the pre-exercise resting value. On the other hand, in the absence of any peak heart rate after the initial increase in the supine position, the increased heart rate reached a steady level without increasing or decreasing simply in a step-increase way. What is important here is that the steady-state heart rate was lower in the upright than in the supine position (71.9 ± 1.2 versus 77.2 ± 1.4 beats/min, $P < 0.05$). In general, heart rate is always higher in the upright than in the supine position, both at rest and during exercise at the same work rates. At the time, cardiac output is evidently lower in the upright than in the supine position. This difference depends largely on the stroke volume, because the increase in heart rate does not fully compensate. In this study, total peripheral resistance was clearly greater in the upright than in the supine position. Importantly, there was no significant difference in mean blood pressure between the two positions during rest or exercise.

4.4.2 Vagal Withdrawal Evoked by Central Command

The immediate rise in heart rate at the onset of exercise in the upright position is due to withdrawal of vagal nerve tone to the heart that would originate from centrally generated command signals, called central command. The central command acts to activate autonomic neurons within the cardiovascular control center located in the brainstem. The increased supine heart rate in a step-like function is thought to be also due to the decreased activity of vagal nerve efferents evoked by the central command. The subsequent decrease in upright heart rate after its peak would be attributable to an increase in vagal nerve activity, a decrease in sympathetic nerve activity, or both. The parasympathetic- and sympathetic-mediated cardiac responses are theoretically and physiologically possible. The potential speed of parasympathetic input effects can be within 1 s. On the contrary, the effect of sympathetic input is relatively slow, taking 10–20 s. Thus, if central command withdraws cardiac vagal nerve traffic, granted that it increases sympathetic nerve traffic at the same time, the initial increase in heart rate and the subsequent decrease over the first 10–20 s would be mediated by vagal withdrawal and vagal reactivation, respectively. The effects of mechanosensory activated sympathetic efferents from working muscles would also not appear for approximately 10–15 s. In support of the above mention, some pharmacological studies have reported that such heart rate responses or tachycardia is blocked by atropine and not by propranolol.

4.4.3 Transiently Fall in Blood Pressure

Here, let us consider the effect of the arterial baroreflex on the control of the circulation. Immediately after the start of exercise, total peripheral resistance decreased sharply in each of the positions. The increase in cardiac output due to the rapid increase in heart rate could not compensate for the decrease in total peripheral resistance, resulting in the transient decrease in mean blood pressure. The subsequent mean blood pressure was, however, maintained at a rest level during exercise in both positions. The blood vessels are compressed during exercise by the contraction of the muscles that surround them. In particular, highly compliant veins are compressed by the muscle, while through the increased blood pressure in these veins, the pressure head from the feet to the heart becomes large. Furthermore, the venous return of one-way flow to the heart is promoted together with the backflow prevention function of the venous valves. On the other hand, the arterial system of relatively low compliance has smaller vascular volume, so that the promotion of venous return causes the arterial blood to be drawn in due to the decrease in venous pressure, leading to the decreased systemic blood pressure. At the time, the transient decrease in blood pressure would occur because compensation for the decreased total peripheral resistance cannot be met in time with delayed increase in cardiac output.

When an increase in cardiac output is completed after an initial transition from rest to exercise, the relationship between total peripheral resistance and cardiac output will lead to a stable arterial pressure. The set point of baroreflex is reset to a new pressure level so as to buffer the error of blood pressure to the target stable level via the reflex control. However, the resetting of the baroreflex would be in transition during the initial phase after the start of exercise. From our results and the others' together, it is conceivable that this complex change in heart rate immediately after the start of exercise in upright posture is not mediated by control of the sinoaortic baroreflexes. A detailed explanation of this mechanism is given in the following section.

4.4.4 Functional Sympatholysis

Reduced total peripheral resistance during exercise in both postures consists of vasodilation in the active muscular tissues. Vasoconstriction in inactive muscles or other inactive organs will also redistribute blood to the working muscles; however, a feedback contribution of muscle chemosensory reflex to sympathetic vasoconstriction in active muscles during low-intensity exercise may be weak as a factor to regulate the circulation. In the supine position, discharge frequency of the cardiopulmonary mechanoreceptor nerves would augment to the medullary cardiovascular control center, which suppresses sympathetic nerve activity. A decrease in tonic sympathetic nerve activity can reduce the normal tonus of the vasculature in the whole body, such as legs, arms, splanchnic organs, and kidneys. With these vasoconstrictions during exercise, an amount of blood is feebly redistributed from inactive tissues to working skeletal muscles. The redistribution of blood with finely adjusted arterial pressure is achieved through the interplay between sympathetic vasoconstrictor outflow and vasodilative substances released from the active muscles. The sympathetic vasoconstriction, adrenergic vasoconstriction, is attenuated locally in the exercise muscle, the response of which is termed "functional sympatholysis." The mechanisms of functional sympatholysis have been extensively studied with regard to ATP, nitric oxide, prostaglandins, potassium channels, and Na^+/K^+-ATPase, which also play potential roles in blood flow regulation during exercise.

The pressure environment in the cardiopulmonary vasculature, where the mechanosensory receptors are located, is affected by posture, which is simply summarized with the arterial baroreceptors in Table 4.2. At rest, the rate of cardiopulmonary mechanoreceptor nerve firing can be higher in the supine position due to the loading of central venous volume than in the upright position. Although total peripheral resistance was lower and cardiac output was higher in the supine than in the upright position, the systemic blood pressure should be controlled at almost the same level in both the positions. In normal physiological states, both at rest and during exercise in the two different positions, higher supine cardiac output than upright cardiac output was achieved by relatively high stroke volume, not by heart rate. In contrast, in the sitting position, increased total peripheral resistance would be due to increase in sympathetic vasoconstrictor tone to the peripheral vessels via reduced activity of cardiopulmonary mechanoreceptor afferents. The difference in total peripheral resistance between the two positions was smaller during exercise than during rest (*bottom* in Fig. 4.4). The reason for this small difference is that the increase in venous return due to the skeletal muscle pump would further reduce the sympathetic-mediated vasoconstriction via the cardiopulmonary receptor inhibitory reflex during upright exercise compared with rest.

Table 4.2 Relative hydrostatic pressure levels

	Upright				Supine		
	Level	CP	Aortic	Carotid	CP	Aortic	Carotid
Rest	High				O	O	O
	Medium	O	O				
	Low			O			
Exercise	High	O	O		O	O	O
	Medium	↑	↑	O			
	Low			↑			

CP cardiopulmonary baroreceptor regions, *Aortic* aortic arch regions, *Carotid* carotid sinus regions

4.4.5 Cardiopulmonary Mechanoreceptor Reflexes

Venous pressure in the foot decreases to about 30 mmHg during locomotion or repeated muscular contractions (see Chap. 5 for details). For this phenomenon, it takes a certain time for blood to move from the peripheral vein to the central vein. As a result, increases in left ventricular blood filling and stroke volume lag behind the initial increase in heart rate at the onset of exercise (lower panel in Fig. 4.3). Even if ventricular contractility is increased, increased stroke volume should be transient with insufficient filling from an unbalance between the cardiac inflow and outflow. Thus, the initial fall in systemic blood pressure, as described above, is probably unrelated to the baroreflexes. The decrease in arterial pressure after its initial decrease is not due to increased sympathetic nerve activity, because the decreased heart rate at the time is considered to be elicited by reactivation of the vagus nerve. The little or no involvement of sympathetic nerves to the vasculature is supported by the fact that total peripheral resistance was not increased with exercise. In comparison to supine heart rate, the different responses of decreased heart rate to total peripheral resistance during upright exercise in this study may be unusual for the physiological function of the cardiovascular system. The parasympathetic and sympathetic mechanisms of the autonomic nervous system usually have opposite reflex effects on arterial baroreflexes; for example, when the heart rate decreases reflexly with the recapture of vagal tone, total peripheral resistance in a steady state, except for a transient, will also decrease with the withdrawal of sympathetic tone.

A role of cardiopulmonary baroreceptors in the reflex control of heart rate in recumbent resting conscious dogs has been known as an increase in heart rate that can be induced by intravenous infusion of saline. The tachycardia is attributed to withdrawal of vagal tone and is called the Bainbridge reflex. In healthy humans, however, variations in cardiac filling pressure and central venous pressure provoked by lower-body negative pressure and by elevation of the legs within the physiological range could not evidently affect sinoatrial node response to cardiopulmonary mechanoreceptor stimulation. Several studies have thus failed to demonstrate the Bainbridge effect in humans under physiological conditions. Therefore, the Bainbridge reflex probably plays little or no physiological role in control of heart rate in humans. Moreover, since there is lack of conclusive evidence for such a reflex tachycardia or bradycardia in humans, caution must be taken to construct evidence of whether cardiopulmonary mechanoreceptors affect heart rate in the pathological heart of organic disease and neuropathy.

Several studies have shown that a transient decrease in heart rate after its initial rise is apparent during the transition from quiet rest to exercise, but ambiguous from light to heavy exercise. The biphasic fluctuation is also clearly seen at the onset of exercise in the upright as compared with the supine posture. In this study, in accordance with these results, the overshooting of heart rate was observed during cycle exercise at light intensity that started from quiet rest only in the upright seated position.

The present study was undertaken to examine the difference in heart rate response to the transition from rest to dynamic leg exercise at light intensity between the upright and supine positions. The cardiopulmonary reflexogenic regions located mainly at one junction of the vena cavae and another junction of the pulmonary veins are richly innervated by sensory receptors whose afferent nerve fibers travel to the central nervous system in the brainstem. The cardiopulmonary mechanosensory afferent nerves of the reflex control play important roles in tonically maintaining and modulating the degree to which arterial baroreflexes elicit the vagal and sympathetic outflows to the heart and peripheral vasculature. The initial rapid increase in heart rate at the onset of exercise is thought to be due to the withdrawal of the vagal efferent nerves to the heart that originates from the central motor command. From what has been discussed, we have hypothesized that the reduction in heart rate after its initial peak during light-intensity dynamic exercise in the upright position is mediated by the vagal tonic discharge of the cardiopulmonary mechanoreceptor reflex. The cardiopulmonary neural afferents modify the autonomic outflow to the sinoatrial node of the heart through the cardiovascular control center in the central nervous system. Therefore, heart rate responses to light-intensity dynamic exercise can be differentiated by different postures. Both central blood volume and cardiac filling vary markedly owing to a gravitational shift of blood volume in the circulation with changing posture.

In this study, on the transition from rest to exercise, the magnitude of the increase in stroke volume was much greater in the upright than in the supine position, in agreement with the results reported by other researchers; however, in general, the absolute values for stroke volume in the upright are always less than those in the supine position. In a human study, cardiovascular responses to graded bicycle exercise have been examined in a supine position during lower-body negative pressure in the absence of hypotension. The hemodynamic responses to supine exercise during nonhypotensive lower-body negative pressure are qualitatively similar to those to upright exercise. That study has found that a steady-state level of heart rate during unloaded cycling with lower-body negative pressure significantly reduces from a resting baseline level, while stroke volume increases above the resting baseline level. In contrast, during controlled supine exercise without lower-body negative pressure, heart rate remained elevated above the resting baseline level, whereas stroke volume was almost unchanged. The increase in stroke volume during exercise in the upright posture in this study may be accompanied by increases in central venous volume (or pressure) and cardiac filling volume (or pressure), which should result from an augmentation of venous return by means of the action of the muscle pump. This increased central venous volume probably loads and discharges the cardiopulmonary mechanoreceptors (Table 4.2). Accordingly, other researchers have previously published the data that light knee-extension exercise performed in upright posture leads to the suppression of muscle sympathetic nerve activity to nonworking muscles.

4.4.6 Venous Return by Skeletal Muscle Pump

The present study has demonstrated that the steady-state heart rate during upright exercise at 30 W or about 20% of peak oxygen uptake was significantly decreased from the initial peak and was then significantly lower than that during supine exercise. Several neural inputs including afferent fibers of the central motor command, mechanoreceptors and chemoreceptors in the muscles, and central and peripheral chemoreceptors are unlikely to be responsible for the difference in heart rate response to the upright and supine exercise, because the subjects in the upright and supine positions perform light exercise at the same workload and have similar oxygen consumptions.

Furthermore, the high-frequency amplitude of R–R interval variability during exercise was greater in the upright than in the supine position. These results suggest that the decreased heart rate after the initial increase during upright exercise is mediated by an increase in cardiac vagal nerve activity indirectly via the cardiopulmonary mechanoreceptor reflex activated by the augmentation of venous return due to the skeletal muscle pump. The supine position would not effectively promote the skeletal muscle pump to move venous blood flow toward the heart, since the facilitative effect under gravity on arterial blood movement cannot be operative. Accordingly, a deliberate study of Folkow et al. has demonstrated that muscle blood flow during exercise is higher in the upright than in the supine position.

4.4.7 Interaction Between Arterial and Cardiopulmonary Baroreflexes

Shown in Fig. 4.3 is the greater stroke volume in the supine position, the cardiac chambers or the thoracic vasculature during supine exercise can be loaded with a substantially larger blood volume than during upright exercise. In this context, the difference in heart rate response to the same exercise between supine and upright positions cannot be fully explained by the reflex modulation of the autonomic outflows to the heart via cardiopulmonary mechanoreceptors. Alternatively, the sensitivity of arterial baroreflexes for the heart rate-arterial pressure response curve at a set point during exercise may be different between the upright and supine positions. The reflex sensitivity, the set-point location, or both may also be modified by the afferent neural input from the cardiopulmonary mechanoreceptor afferents to the central nervous system. Numerous investigators have provided indirect data to support or refute this assumption that the cardiopulmonary afferent pathway could modulate the arterial baroreflex control of heart rate. In all of these studies, the influences of cardiopulmonary baroreceptor nerve afferents on the reflex control of heart rate were examined under the conditions of lower-body negative pressure or central venous infusion in the supine position. In these studies, the central venous expansion may inappropriately stimulate the cardiopulmonary mechanoreceptors, because the central blood volume was already elevated in the supine position before the infusion. By contrast, in the upright position, when an acute increase in central blood volume appropriately stimulates the cardiopulmonary mechanoreceptors, for example, during head-out water immersion or during blood infusion, heart rates decrease substantially below the prestimulation levels; however, arterial baroreceptor loading, which is involved in such reflex decrease in heart rate, is still functioning during head-out water immersion. In our study, it is evident that there was no difference in systemic arterial pressure during exercise between the upright and supine positions. Thus, the cardiopulmonary mechanosensory neural input to the central nervous system would cause the arterial baroreceptor to modulate the control of heart rate in the upright position. Decisively, the cardiopulmonary mechanosensory input is a potential factor, underlying the mechanism that reduces heart rate in response to upright exercise at light intensity.

4.4.8 Extravascular Fluid Homeostasis in Muscle Tissues

In the blood vessels of the lower extremities, the difference between intravascular and extravascular pressure (transmural pressure) is larger in the upright than in the supine position through the gradient of hydrostatic pressure under gravity. In particular, at arteriolar levels in the upright position, the autoregulation of blood flow control is thought to work predominantly with the wall of blood vessels that will be tonically loaded by the transmural pressure. During exercise, the autoregulation and muscle pump help the postcapillary venules to collect tissue fluid in a phase of decreased venous pressure during the periods of muscular relaxation. At the same time, the action of muscle pump also plays an important role in maintaining the extravascular fluid homeostasis of the muscular tissue, since it effectively promotes lymph flow in lymphatic vessels in the exercising muscle.

4.4.9 Physiological Significance and Clinical Implication

The present study simply examined a short-term response of heart rate to light exercise in upright posture. Still, we should also consider the physiological significance of decreased heart rate and increased vagal nerve activity. From the results of our and others' studies, it is possible to argue that stimulation of cardiopulmonary baroreceptors due to increased venous return during upright dynamic exercise at light intensity induces increased parasympathetic nerve activity and decreased sympathetic nerve activity. If an increase in cardiac vagal nerve activity is induced during light exercise in the upright position as mentioned above, the vagal tone maintained reflexly can play an important role in protection against sudden cardiac arrest, especially during rehabilitative training for patients with cardiovascular diseases and during exercise training for healthy elderly. There is also a report that efferent cardiac vagal nerve activation can effectively prevent in part ventricular fibrillation caused by acute myocardial ischemia. Our hypothesis for exercise-induced cardiac vagal tone through the cardiopulmonary mechanoreflexes has been supported in part by the findings in the literature that increased parasympathetic tone as well as decreased sympathetic tone caused by training results in significant protection from sudden cardiac death. Therefore, the vagal activation of cardiopulmonary mechanoreflexes is likely to play an important role in antagonizing both fatal arrhythmia and sympathetic-mediated arrhythmia, particularly during a warm-up and a cool-down gentle exercise in the upright position.

4.4.10 Limitations

In this study, the subjects had to perform extra work at the start of exercise, in addition to the workload of 30 W, in order to overcome the inertial force of the static flywheel. In an elegant study of Sietsema et al., an electric motor was used to drive the flywheel before the start of exercise on a cycle ergometer. The response of heart rate to the assisted exercise starting from rest showed an initial sharp increase and the subsequent decrease, which is consistent with the results in this study. Thus, the initial extra work to overcome inertial force of flywheel should not notably affect our conclusions.

4.5 Summary

A decrease in heart rate after its initial rise during the transition from rest to light exercise were studied in eight untrained male subjects, who performed cycle exercise at 20% of peak oxygen uptake in upright (sitting) and supine positions. Heart rate increased to a peak of 86.3 ± 1.4 (mean ± SE) beats/min at 25 ± 1 s after the start of upright exercise and decreased to just above the resting level thereafter. In contrast, heart rate during supine exercise from rest had no corresponding peak, but increased in a step-like manner. The mean heart rate at the sixth minute during exercise was significantly less in the upright than in the supine position (71.9 ± 1.2 vs. 77.2 ± 1.4 beats/min). Upright stroke volume (by impedance cardiography) increased significantly by about 20% from rest to exercise with the help of the muscle pump. Conversely, supine stroke volume was unchanged with probably already elevated resting central blood volume; at the time, changes in stroke volume were used to presume relative changes in central blood volume. The muscle pump during upright exercise can augment the central blood volume, thereby activating the cardiopulmonary mechanoreceptors. The high-frequency amplitude of R–R interval variability, an index of cardiac vagal tone, in power spectral analysis was significantly less in the upright than in the supine position at rest, but vice versa during exercise. Mean arterial blood pressures at rest and during exercise were not significantly different between the two positions. The heart rate and cardiac output increased greatly, but the mean blood pressure showed little or no change (except the initial drop) in the upright and supine positions. The fall in total peripheral resistance could allow the perfusion of the active muscles without a rise in arterial pressure. These results suggest that the initial rise and subsequent reduction in heart rate during the transition from rest to exercise in the upright position are associated with withdrawal of vagal nerve efferents mediated by the central command and the subsequent vagal reactivation mediated by cardiopulmonary baroreflex, respectively.

Further Reading

Takahashi T, Ashikawa K, Okada A, Takeshima N. Hypothesis: a role of cardiac receptor nerve afferent in reflex control of heart rate during light exercise in upright humans. Ther Res. 2003;24:1583–93.

Abboud FM, Eckberg DL, Johannsen UJ, Mark AL. Carotid and cardiopulmonary and forearm vascular resistance during venous pooling in man. J Physiol. 1979;286:173–84.

Bainbridge FA. The influence of venous filling upon the rate of the heart rate. J Physiol. 1915;50:65–84.

Bayliss WM. On the local reactions of the arterial wall to changes in internal pressure. J Physiol. 1902;28:220–31.

Bevegård S, Holmgren A, Jonsson B. The effect of body position on the circulation at rest and during exercise, with special reference to the influence on the stroke volume. Acta Physiol Scand. 1960;49:279–98.

Billman GE, Dickery DT, Teoh KK, Stone HL. Effects of central venous blood volume shifts on arterial baroreflex control of heart rate. Am J Phys. 1981;241:H571–5.

Billman GE, Schwartz PJ, Srone HL. The effects of daily exercise on susceptibility to sudden cardiac death. Circulation. 1984;69:1182–9.

Bishop VS, Malliani A, Thorén. Cardiac mechanoreceptors. In: Shepherd JT, Abboud FM, Geiger SR, editors. Handbook of physiology. The cardiovascular system, vol. 3. Bethesda: American Physiological Society; 1983. p. 497–555.

Blomqvist CG, Stone HL. Cardiovascular adjustments to gravitational stress. In: Shepherd JT, Abboud FM, Geiger SR, editors. Handbook of physiology. The cardiovascular system, vol. 3. Bethesda: American Physiological Society; 1983. p. 1025–63.

Borst C, Wieling W, Van Brederrode JFM, Hond A, De Rijk LG, Dunning AJ. Mechanisms of initial heart rate response to postural change. Am J Phys. 1982;243:H676–81.

Brown AM. Cardiac reflexes. In: Berne RM, Sperelakis N, Geiger SR, editors. Handbook of physiology. The cardiovascular system, vol. 1. Bethesda: American Physiological Society; 1979. p. 677–89.

Crecelius AR, Kirby BS, Hearon CM, Luckasen GJ, Larson DG, Dinenno FA. Contracting human skeletal muscle maintains the ability to blunt α_1-adrenergic vasoconstriction during K_{IR} channel and Na^+/K^+-ATPase inhibition. J Physiol. 2015;593:2735–51.

Eiken O. Responses to dynamic leg exercise in man as influenced by changes in muscle perfusion pressure. Acta Physiol Scand Suppl. 1987;131:1–37.

Epstein M. Renal effects of head-out water immersion in man: implications for an understanding of volume homeostasis. Physiol Rev. 1978;58:529–81.

Fagraeus L, Linnarsson D. Autonomic origin of heart rate fluctuations at the onset of muscular exercise. J Appl Physiol. 1976;40:679–82.

Flamm SD, Taki J, Moore R, Lewis SF, Keech F, Maltais F, Ahmad M, Callahan R, Dragotakes S, Alpert N, Strauss W. Redistribution of regional and organ blood volume and effect on cardiac function in relation to upright exercise intensity in health human subjects. Circulation. 1990;81:1550–9.

Folkow B, Haglund U, Jodal M, Lundgren O. Blood flow in the calf muscle of man during heavy rhythmic exercise. Acta Physiol Scand. 1971;81:157–63.

Fortney SM, Wenger CB, Bove JR, Nadel ER. Effect of blood volume on forearm venous and cardiac stroke volume during exercise. J Appl Physiol. 1983;55:884–90.

Freyschuss U. Cardiovascular adjustment to somatomotor activation. Acta Physiol Scand Suppl. 1970;342:1–63.

Gauer OH, Thron HL. Postural changes in the circulation. In: Hamilton WF, Dow P, editors. Handbook of physiology. Circulation, vol. 3. Washington: American Physiological Society; 1965. p. 2409–39.

Hainsworth R. Reflexes from the heart. Physiol Rev. 1991;71:617–58.

Hearon CM, Richards JC, Racine ML, Luckasen GJ, Larson DG, Joyner MJ, Dinenno FA. Sympatholytic effect of intravascular ATP is independent on nitric oxide, prostaglandins, Na^+/K^+-ATP and K_{IR} channels in humans. J Physiol. 2017;595:5175–90.

Heymans C, Neil E. Reflexogenic areas of the cardiovascular system. London: J & A Churchill Ltd.; 1958.

Hohanty PK, Thames MD, Arrowood JA, Sowers JR, McNamara C, Szentpetery S. Impairment of cardiopulmonary baroreflex after cardiac transplantation in humans. Circulation. 1987;75:914–21.

Imholz BP, Settels JJ, Van Der Meiracker AH, Wesseling KH, Wieling W. Non-invasive continuous finger blood pressure measurement during orthostatic stress compared to intra-arterial pressure. Cardiovasc Res. 1990;24:214–21.

Johnson JM, Rowell LB, Niederberger M, Eisman MM. Human splanchnic and forearm vasoconstrictor responses to reductions of right atrial and aortic pressures. Circ Res. 1974;34:515–24.

Katona PG, Jih F. Respiratory sinus arrhythmia: non-invasive measure of parasympathetic cardiac control. J Appl Physiol. 1975;39:801–5.

Kubicek WG, Karnegis JM, Patterson RP, Witsoe DA, Mattson RH. Development and evaluation of an impedance cardiac output system. Aerosp Med. 1966;37:1280–12.

Maciel BC, Gallo L Jr, Marin Neto JA, Lima Filho EC, Martins LEB. Autonomic nervous control of the heart rate during dynamic exercise in normal man. Clin Sci. 1986;71:457–60.

Maciel BC, Gallo L Jr, Marin Neto JA, Martins LE. Autonomic nervous control of the heart rate during isometric exercise in normal man. Pflugers Arch. 1987;408:173–7.

Maehlum S, Grandmontagne M, Newsholme EA, Sejersted OM. Magnitude and duration of excess postexercise oxygen consumption in healthy young subjects. Metabolism. 1986;35:425–9.

Mark AL, Mancia G. Cardiopulmonary baroreflexs in humans. In: Shepherd JT, Abboud FM, Geiger SR, editors. Handbook of Physiology. The cardiovascular system, vol. 3. Bethesda: American Physiological Society; 1983. p. 795–813.

Miyamoto Y, Tamura T, Hiura T, Nakamura T, Higuchi J, Mikami T. The dynamic response of the cardiopulmonary parameters to passive head-up tilt. Jpn J Physiol. 1982;32:245–58.

Miyamoto Y, Higuchi J, Abe Y, Hiura T, Nakazono Y, Mikami T. Dynamics of cardiac output and systolic time intervals in supine and upright exercise. J Appl Physiol. 1983;55:1674–81.

Öberg B, White S. Circulatory effects of interruption and stimulation of cardiac vagal afferents. Acta Physiol Scand. 1970;80:383–94.

Paterson DJ. Antiarrhythmic mechanisms during exercise. J Appl Physiol. 1996;80:1853–62.

Perini Z, Milesi S, Biancardi L, Pendergast DR, Veicsteinas A. Heart rate variability in exercising humans: effect of water immersion. Eur J Appl Physiol. 1998;77:326–32.

Poliner LR, Dehmer GJ, Lewis SE, Parkey RW, Blomquist CG, Willerson JT. Left ventricular performance in normal subjects: a comparison of the responses to exercise in the upright and supine positions. Circulation. 1980;62:528–34.

Ray CA, Rea RF, Clary MP, Mark AL. Muscle sympathetic nerve responses to dynamic one-legged exercise: effect of body posture. Am J Phys. 1993;264:H1–7.

Risch WD, Koubenec H-J, Beckmann U, Lange S, Gauer OH. The effect of graded immersion on heart volume, central venous pressure, pulmonary blood distribution, and heart rate in man. Pflugers Arch. 1978;374:115–8.
Rowell LB. Human circulation regulation during physical stress. New York: Oxford University Press; 1986.
Rowell LB. Human cardiovascular control. New York: Oxford University Press; 1993.
Sawada Y, Ohtomo N, Tanaka Y, Tanaka G, Yamakoshi K, Terachi S, Shimamoto K, Nakagawa M, Satoh S, Kuroda S, Imurai O. New technique for time series analysis combining the maximum entropy method and non-linear least squares method: its value in heart rate variability analysis. Med Biol Eng Comput. 1997;35:318–22.
Sietsema KE, Daly JA, Wasserman K. Early dynamics of O_2 uptake and heart rate as affected by exercise work rate. J Appl Physiol. 1989;67:2535–3541.
Smith JJ. Circulatory response to the upright posture. Boca Raton: CRC Press; 1990.
Starling EH. The fluid of the body. London: Archibald Constable; 1909.
Stegall HF. Muscle pump in the dependent leg. Circ Res. 1966;19:180–90.
Takahashi T, Miyamoto Y. Influence of light physical activity on cardiac responses during recovery from exercise in humans. Eur J Appl Physiol. 1988;77:305–11.
Takahashi T, Okada A, Saitoh T, Hayano J, Miyamoto Y. Difference in human cardiovascular response between upright and supine recovery from upright cycle exercise. Eur J Appl Physiol. 2000;81:233–9.
Task Force of the European Society of Cardiology and the North American Society of Pacing and Electrophysiology. Heart rate variability. Standards of measurement, physical interpretation, and clinical use. Circulation. 1996;93:1043–65.
Vanoli E, De Ferrari GM, Stramba-Badiale M, Hull SS, Foreman RD, Schwartz PJ. Vagal stimulation and prevention of sudden death in conscious dogs with a healed myocardial infarction. Circ Res. 1991;68:1471–81.
Vissing SF, Scherrer U, Victor RG. Relation between sympathetic outflow and vascular resistance in the calf during perturbations in central venous pressure. Circ Res. 1989;65:1710–7.
Weiler-Ravell D, Cooper DM, Whipp BJ, Wasserman K. Control of breathing at the start of exercise as influenced by posture. J Appl Physiol. 1983;55:1460–6.
Whipp BJ, Ward SA, Lamarra N, Davis JA, Wasserman K. Parameters of ventilatory and gas exchange dynamics during exercise. J Appl Physiol. 1982;52:1506–13.
Yoffey JM, Courtice FC. Lymphatics, lymph and lymphoid tissue. London: Edward Arnold Publishers Ltd; 1956.

Chapter 5
Skeletal Muscle Pump and Body Posture

5.1 Introduction

Exercise exerts an antihypertensive effect in normotensive and hypertensive subjects, regardless of whether the exercise is performed in an upright or a supine position. At the time, postural changes produce alterations in hemodynamics which are disturbed by the force of gravity. Hence, characterizations of cardiovascular functions with various reflexes, vasoactive substances, and hormones during exercise and recovery are required for a basal understanding of the influence of postural difference on control of blood pressure.

Objective assessment of the cardiovascular system during recovery from exercise has created considerable scientific and clinical interest. For example, active-recovery exercise plays important roles in facilitating the restoration of vagal nerve outflow and the venous return of blood to the heart and in reducing the risks of postexercise orthostatic hypotension and sudden cardiac arrest. In addition, ventricular stroke volume during active recovery (light-load cycling) in an upright position is maintained at the same level as that during exercise, whereas that cannot be during inactive recovery (complete rest). The maintenance of this stroke volume conforms with the Frank-Starling mechanism, the background of which is as follows: venous return and cardiac filling are increased by skeletal muscle contraction, the so-called "muscle pump." On the other hand, the effects of postexercise activity and muscle pump on cardiovascular responses during active recovery from exercise in a supine position remain relatively uncharacterized. To learn the characteristics of the circulatory system in the upright position can be very useful in understanding those in the supine position, and vice versa.

The present chapter is concerned with differences in the postexercise regulation of the cardiovascular system between inactive recovery and active recovery from moderate cycle exercise in an upright sitting position and a supine lying position. In particular, we clarify the effects of the muscle pump and body position on the

T. Takahashi, *The Cardiovascular System During Exercise and Recovery*,
https://doi.org/10.1007/978-981-99-8486-2_5

control of arterial blood pressure during postexercise recovery. For these interests, cardiovascular variables are compared in response to the following two modes of recovery in both positions: (1) inactive recovery, during which subjects quietly rest after exercise; and (2) active recovery, during which subjects perform light exercise.

This study evaluates the hypothesis that central blood volume and cardiac filling pressure would remain unchanged from their pre-exercise resting levels during active and inactive recovery in supine posture, regardless of whether the muscle pump works or not. Some studies have examined that hardly increased either stroke volume or central venous pressure from rest to exercise in supine posture, whereas those significantly increased in upright posture; these results have provided support for the hypothesis. In a comparison between the upright and supine positions, the difference in blood distribution under gravity will reveal information as to how the circulatory system controls blood pressure and flow. Also, as for a comparison of active and inactive recovery, the centripetal flow of blood in the lower limbs promoted by the muscle pump indicates a way of the circulatory system controlling the arterial blood pressure. Resting and exercising values for cardiovascular variables (except mean blood pressure) considerably differ between the upright and supine positions, as shown in the previous chapters. Clarifying essential differences in hemodynamics between inactive and active recovery and between the two positions can provide us important information to consider the cardiovascular mechanism underlying the control of systemic arterial blood pressure during recovery from exercise.

Therefore, this chapter describes an overview of how time courses of dynamic responses of the cardiovascular system in such components as heart rate, stroke volume, cardiac output, blood pressure, and systemic vascular conductance are different during active and inactive recovery in the different body positions. It will also be discussed that the integrated interplay between the heart and the vasculature enables the cardiovascular system to control the arterial blood pressure.

5.2 Materials and Methods

5.2.1 Human Subjects

Seven healthy young male volunteers participated in this study. The physical characteristics of the subjects with a mean age of 22.4 ± 2.1 years (mean ± SD) were as follows: a mean height, 1.70 ± 0.06 m; a mean body mass, 61.1 ± 9.1 kg; and a mean peak oxygen uptake, 2.73 ± 0.40 L/min. A cardiologist examined them, and they had no signs of cardiovascular disease. All the subjects were informed about the procedures, risks, and benefits of the study, and then provided written consent forms before the study. None of them was engaged in any regular physical activity or involved in competitive sports. All experimental protocols and procedures conformed to the Declaration of Helsinki. The experimental protocols were also approved by the Health Research Ethics Board at the university.

5.2.2 Experimental Protocols

On a preliminary visit, the subjects underwent a cycle exercise test for determination of peak oxygen uptake on a stationary cycle ergometer with an electromagnetically braked resistance (Corival 300; Lode, The Netherlands) in an upright seated position. They underwent a ramp-exercise test with an incremental work rate of 15 W/min until they could not continue to cycle exercise any more. This end point termed "exhaustion" was taken as a point at which they were unable to maintain a pedaling rate of at least 50 revolutions/min.

The subjects were allowed some practice to become accustomed to the two types of cycle exercise; specifically, one exercise was performed in an upright sitting position on the ergometer and the other was performed in a supine lying position on a bed behind the ergometer. These practices were completed a week before the commencement of the main exercise tests.

The subjects performed inactive and active recovery from exercise separately in the upright position (Fig. 5.1). They also performed inactive and active recovery from exercise separately in the supine position. The tests consisted of rest for 5 min, cycle exercise for 10 min, and then recovery for 5 min. The intensity of stimulus exercise was set constant at 60% of peak oxygen uptake and the intensity of active-recovery exercise was set at 20% of peak oxygen uptake. Both of the body positions were maintained from the beginning of rest until the end of recovery period. A pedaling frequency of 60 revolutions/min was maintained throughout the exercise and active recovery with the aid of an audible metronome. The pedaling rate was

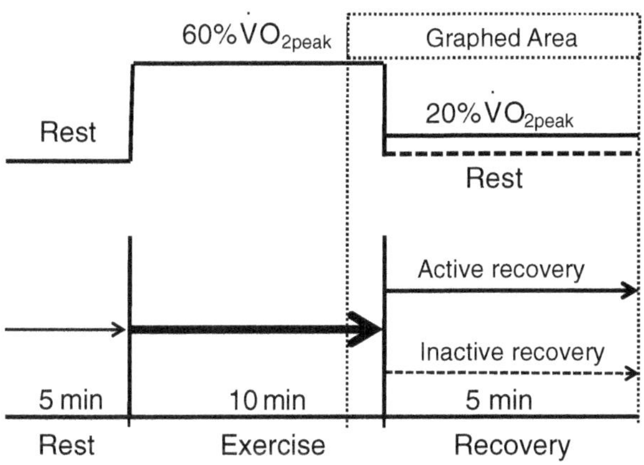

Fig. 5.1 Schematic representation of exercise protocols. Exercise test consisted of rest for 5 min, cycle exercise for 10 min at a relative intensity of 60% of peak oxygen uptake, and recovery for 5 min in either upright sitting position or supine lying position. Each position was maintained throughout the experiments. During recovery period, the following two modes were adopted: (1) no leg movement (inactive recovery), and (2) voluntary leg movement (active recovery) at a relative intensity of 20% of peak oxygen uptake. The cardiovascular variables in the graphed area surrounded by dotted lines are shown in the resulting graphs in the texts

monitored with an inductive revolution meter fitted to the ergometer. The subjects placed their feet on the footrest near the flywheel during rest periods in each body position; importantly, no movement of lower limbs was effective in eliminating a drainage effect of their muscle pumping action on venous return to the heart.

All the experiments were initiated at 10:00 in each subject. The order of trials (e.g., upright vs. supine) was randomly assigned. The subjects performed each test twice for the four types of recovery: active recovery in the upright and supine positions vs. inactive recovery in both the positions. They conducted a trial only once a week to avoid the effects of exercise training on the cardiorespiratory capacity. They were also instructed to abstain from any beverages containing either caffeine or alcohol from 21:00 on the night before exercise trials and to abstain from vigorous exercise for 36 h before trials. They had finished eating a small breakfast at least 3 h before the tests. All the experiments were conducted at a room temperature of 21–22 °C.

5.2.3 Cardiovascular Measurements

Heart rate (*HR* in beats per minute) was calculated from R–R intervals monitored by electrocardiography. Cardiac output (*CO* in liters per minute) was determined by impedance cardiography (ICG). An impedance cardiograph of a standard constant-current-type unit (RGA-5, Nihon Koden, Tokyo) was used to monitor transthoracic impedance with four silver tape electrodes, one pair of which were placed at the neck and the other pair were placed at the torso. Ventricular stroke volume (*SV*, in milliliters per beat) monitored by ICG was calculated from the Kubicek standard equation. A more detailed description of the computerized measurement system has been given in Chap. 1.

Artifacts incidental to respiration and body motion on impedance signals were removed arithmetically by averaging ICG waveforms over 10 cardiac cycles. The validity of stroke volume measured with the measurement system was confirmed under various experimental conditions, including rest, exercise, and postural change in several studies. Although absolute values for impedance-derived stroke volume remain incompletely characterized, there has been a general agreement that the ICG is suitable for estimating relative changes in stroke volume during exercise and during postural change. In particular, the reproducibility of estimates of stroke volume for individuals was crucial in comparison to changes in stroke volume between active and inactive recovery in the same position. Special caution was thus taken to locate the electrodes of the ICG at the same position every time, because the stroke volume measured is a function of the distance between the two inside electrodes.

Arterial blood pressure (in millimeters of mercury, mmHg) was measured noninvasively with a pneumatic finger cuff (2300 Finapres, Omeda, Tokyo). The finger cuff was held at heart level to avoid variations in measurement due to changes in hydrostatic pressure throughout all the experiments. Mean arterial blood pressure (*MBP*) was calculated as diastolic blood pressure (*DBP*) plus one-third of the pulse pressure (the difference between the systolic and diastolic pressures). The validity of finger blood pressure recorded has been previously described in several studies published elsewhere. The features of the blood pressure measurement system are also detailed in the Sect. 2.2.4 of Chap. 2.

The systemic vascular conductance (*SVC*, in liters per minute per milliliters of mercury) was computed as cardiac output divided by mean arterial blood pressure; by definition, the systemic vascular conductance is the reciprocal of total peripheral vascular resistance.

Myocardial oxygen uptake (MVO_2) is the amount of oxygen consumed by the left ventricular myocardium of the heart. Myocardial oxygen uptake can be reasonably estimated as the so-called double product, systolic arterial blood pressure (*SBP*) and heart rate (*HR*). It can be given by the equation,

$$MVO_2\left[\mathrm{mL\,O_2}/(100\,\mathrm{g\,LV})/\mathrm{min}\right] = 0.17\left(\mathrm{HR}\times\mathrm{SBP}\times 10^{-2}\right) - 5.31$$

where LV is the left ventricular myocardium. This formula has been proved to be a useful index of myocardial oxygen consumption during cycle exercise in upright posture; however, the effectiveness of the formula used in supine posture remains unknown. Therefore, the results of this index are for reference only; this issue will receive brief consideration in the Sect. 5.4.

5.2.4 *Data Analysis*

Cardiovascular data obtained from two repetitions of each test for four different runs were subjected to Lagrange interpolation to arrange them at 5-s intervals. The two arranged data for a variable were then averaged to yield a single set of regularly interpolated data for each subject. Group mean values were obtained from averaging the data sets of seven subjects. Consecutive 10-s averages for all variables were computed to compare the time courses during the period of recovery. Grand mean values for all variables in steady states were determined from the last minute of averaged data at rest before exercise, during exercise, and during recovery.

5.2.5 *Statistical Analysis*

One-way analysis of variance for repeated measures was used to assess differences in mean values during the periods of rest, exercise, and recovery in each of the four experiments. When a significant *F* ratio was detected, Scheffé multiple comparison post hoc test was used to identify significant differences. Two-way analysis of variance for repeated measures analyzed differences in the pattern of changes in variables between active and inactive recovery experiments in each position and between upright and supine positions in each of the active and inactive recovery experiments. The differences in mean values between rest, exercise, and recovery were also identified by use of two-sided paired *t* test as a post hoc test. Differences were considered significant for all statistical analyses at $P < 0.05$. All values are expressed as mean ± standard error (SE).

5.3 Results

5.3.1 *Rest and Exercise*

Table 5.1 summarizes the mean values for cardiovascular variables obtained from seven subjects, who performed 5-min rests, 10-min bouts of exercise at 60% of peak oxygen uptake, and then either of the 5-min inactive recovery or the 5-min active recovery at 20% of peak oxygen uptake in each of the upright and supine positions. There were no significant differences in mean value at rest or during exercise between inactive and active recovery experiments in either position. All of the cardiovascular variables increased significantly (each $P < 0.05$) from rest to exercise, except for a slight, but insignificant, decrease in the supine stroke volume.

5.3.2 *Recovery*

The upper panel in Fig. 5.2 presents the time courses of heart rate during inactive and active recovery from exercise in the upright position (*left*) and in the supine position (*right*). Heart rates during inactive and active recovery had initially rapid decreases followed by gradual declines in both positions. The time course of upright heart rate during active recovery was similar ($P = 0.370$) to that during inactive recovery, whereas the time course of supine heart rate during active recovery was significantly ($P < 0.05$) higher than that during inactive recovery. Interestingly, the time courses of heart rate during active recovery were almost identical between the upright and supine positions; as indicated, the mean heart rates at the fifth minute were 90.5 ± 3.8 and 89.1 ± 2.3 beats/min, respectively (Table 5.1).

The upright stroke volume during inactive recovery decreased significantly ($P < 0.05$) from the level at exercise to the baseline level at rest, whereas during active recovery it remained at a level similar to the exercise level (*lower panel* in Fig. 5.2). The mean upright stroke volume at the fifth minute was significantly ($P < 0.05$) greater during active than during inactive recovery (Table 5.1). In contrast, the supine stroke volumes during inactive and active recovery showed slight, but insignificant increases and then remained around the pre-exercise resting levels. There were no significant differences in the time course of stroke volume or in mean stroke volume at the fifth minute between inactive and active recovery in the supine position. Parenthetically, supine stroke volume during exercise before either recovery was slightly, but insignificantly decreased from the resting levels. The stroke volumes at rest and during inactive and active recovery were significantly (each $P < 0.05$) greater in the supine than in the upright position.

Table 5.1 Cardiovascular variables before, during, and after exercise in upright and supine positions

		Rest	Exercise	Recovery
HR **(beats·min^{-1})**				
Upright	*Inactive*	76.4 ± 3.2	121.9 ± 6.1 *	89.8 ± 4.8 *
	Active	78.2 ± 2.9	122.3 ± 5.6 *	90.5 ± 3.8 *
Supine	*Inactive*	63.9 ± 2.5 ‡	114.2 ± 3.7 *	72.7 ± 2.9 * ‡
	Active	62.8 ± 2.3 ‡	116.6 ± 3.7 *	89.1 ± 2.3 *†
SV **(mL)**				
Upright	*Inactive*	75.4 ± 2.9	93.2 ± 9.3 *	71.6 ± 5.6
	Active	70.0 ± 4.0	93.0 ± 6.9 *	92.6 ± 4.7 *†
Supine	*Inactive*	116.5 ± 2.6 ‡	102.9 ± 8.4	123.6 ± 6.3 ‡
	Active	124.6 ± 4.9 ‡	105.2 ± 7.1	115.9 ± 6.8 ‡
CO **(L·min^{-1})**				
Upright	*Inactive*	5.7 ± 0.3	11.1 ± 0.8 *	6.3 ± 0.3
	Active	5.4 ± 0.3	11.2 ± 0.6 *	8.3 ± 0.4 *†
Supine	*Inactive*	7.4 ± 0.2 ‡	11.7 ± 1.0 *	8.9 ± 0.3 ‡
	Active	7.8 ± 0.3 ‡	12.2 ± 0.7 *	10.3 ± 0.5 *†‡
MBP **(mmHg)**				
Upright	*Inactive*	92.0 ± 4.3	104.4 ± 4.8 *	91.8 ± 4.9
	Active	89.8 ± 3.9	104.5 ± 4.2 *	95.0 ± 5.3
Supine	*Inactive*	80.3 ± 3.4 ‡	100.9 ± 3.8 *	83.1 ± 4.1 ‡
	Active	81.0 ± 2.9 ‡	101.9 ± 3.7 *	84.1 ± 2.6 ‡
SVC **(L·min^{-1}·mmHg^{-1})**				
Upright	*Inactive*	0.063 ± 0.002	0.108 ± 0.010 *	0.070 ± 0.005
	Active	0.061 ± 0.003	0.109 ± 0.008 *	0.089 ± 0.006 *†
Supine	*Inactive*	0.093 ± 0.004 ‡	0.116 ± 0.009 *	0.109 ± 0.007 ‡
	Active	0.098 ± 0.008 ‡	0.120 ± 0.007 *	0.123 ± 0.006 * ‡
MVO_2 **[mL·(100 g LV)$^{-1}$·min^{-1}]**				
Upright	*Inactive*	10.0 ± 1.2	27.4 ± 2.5 *	13.6 ± 1.6
	Active	11.4 ± 1.1	28.7 ± 2.2 *	16.5 ± 1.5 *
Supine	*Inactive*	7.3 ± 0.9	25.6 ± 2.2 *	9.9 ± 1.1
	Active	7.1 ± 0.5	26.2 ± 2.2 *	13.4 ± 0.7 *†

Values are presented as mean ± SE for seven subjects. *HR* heart rate, *SV* stroke volume, *CO* cardiac output, *MBP* mean arterial blood pressure, *SVC* systemic vascular conductance, *MVO_2* left ventricular myocardial oxygen uptake

* $P < 0.05$, significantly different from rest

† $P < 0.05$, significantly different from upright for the same mode of recovery. Adapted from Takahashi et al., reproduced by permission

‡ $P < 0.05$, significantly different from inactive recovery in the same position

The upright cardiac outputs during inactive and active recovery showed rapid decreases during the first 2 min and gradual decreases thereafter (Fig. 5.3a); however, the time course was significantly ($P < 0.05$) higher during active than during

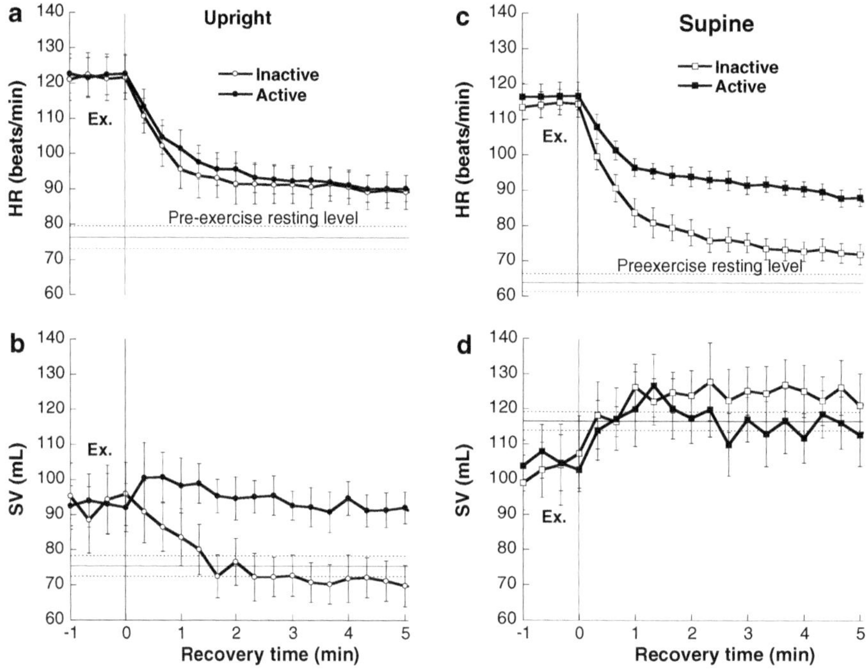

Fig. 5.2 Heart rate (HR, **a** and **c**) and stroke volume (SV, **b** and **d**) during 10-min exercise at 60% of peak oxygen uptake and during inactive recovery and active recovery at 20% of peak oxygen uptake in upright position (left) and supine position (right). Data (mean ± SE) are consecutive 10-s averages obtained from seven subjects for each of the inactive- and active-recovery experiments in both positions. Vertical line (at $t = 0$) indicates the start of recovery. Horizontal thin and dotted lines represent pre-exercise resting levels as mean and ±SE, respectively. Adapted from Takahashi et al., reproduced by permission

inactive recovery. The supine cardiac outputs during inactive and active recovery showed relatively slow, linear decreases (Fig. 5.3d); however, the time course was also significantly ($P < 0.05$) higher during active than during inactive recovery. The higher cardiac output during active recovery in the upright position was due mainly to larger stroke volume, whereas that during active recovery in the supine position was due mainly to higher heart rate. Parenthetically, each of the inactive-recovery and active-recovery cardiac outputs was significantly ($P < 0.05$) greater in the supine than in the upright position (Table 5.1). Notably, the difference in cardiac output for each of the inactive and active recovery between the two positions was largely due to the difference in stroke volume.

The upright systemic vascular conductance during inactive recovery showed a rapid decrease during the first 2 min and a gradual decrease thereafter, whereas that during active recovery showed a gradual decrease after an initially transient increase (Fig. 5.3b). The upright systemic vascular conductance at the fifth minute was significantly ($P < 0.05$) greater during active than during inactive recovery. The supine systemic vascular conductance during inactive and active recovery showed prompt increases and then relatively slow, linear decreases (Fig. 5.3e). The mean difference in supine systemic vascular conductance at the fifth minute between inactive and

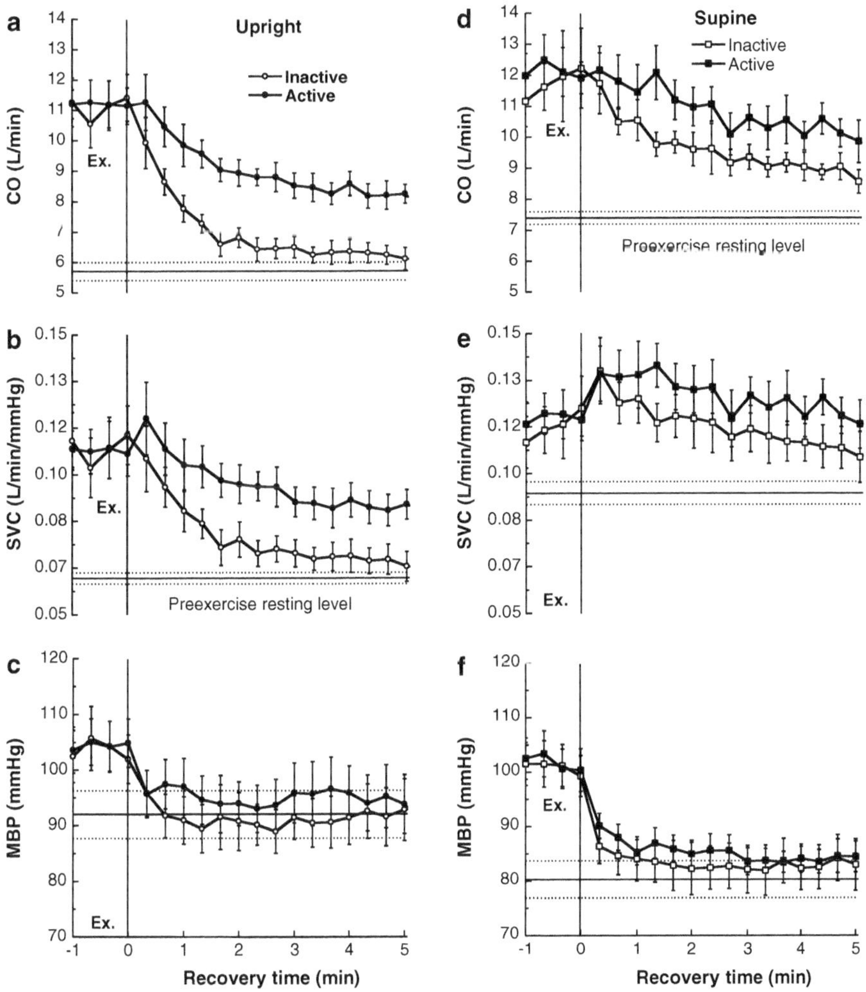

Fig. 5.3 Cardiac output (CO, **a** and **d**), systemic vascular conductance (SVC, **b** and **e**), and mean blood pressure (MBP, **c** and **f**) during 10-min exercise at 60% of peak oxygen uptake and during inactive recovery and active recovery at 20% of peak oxygen uptake in upright position (left) and supine position (right). Data (mean ± SE) are consecutive 10-s averages obtained from seven subjects for each of the inactive- and active-recovery experiments in both positions. Vertical line (at $t = 0$) indicates the start of recovery. Horizontal thin and dotted lines represent pre-exercise resting levels as mean and ±SE, respectively. Adapted from Takahashi et al., reproduced by permission

active recovery was not significant, because three out of seven subjects had opposite results. The systemic vascular conductance at rest and during inactive and active recovery was significantly (each $P < 0.05$) greater in the supine than in the upright position (Table 5.1).

During inactive and active recovery, mean arterial blood pressures in upright and supine positions promptly decreased within the first 2 min almost to the pre-exercise resting levels in the corresponding positions. There was no significant difference in the time course of mean blood pressure between inactive and active recovery in either the upright or supine position. Particularly, the grand mean blood pressures at the fifth minute during inactive and active recovery were significantly ($P < 0.05$) greater in the upright than in the supine position (Table 5.1).

The upper panel in Fig. 5.4 shows that the myocardial oxygen uptakes during inactive and active recovery had rapid decreases during the first minute and gradual decreases thereafter in the upright and supine positions. The upright myocardial oxygen uptake was greater during active than during inactive recovery; however, the difference was insignificant at the fifth minute (Table 5.1). The supine myocardial oxygen uptake was greater during active than during inactive recovery; the difference was statistically significant ($P < 0.05$) at the fifth minute.

The lower panel in Fig. 5.4 demonstrates the time courses of systolic and diastolic arterial blood pressures during inactive and active recovery in the upright and supine positions. The upright systolic blood pressure was higher during active than during inactive recovery, between which there was a significant ($P < 0.05$) difference in mean value at the fifth minute. In contrast, the supine systolic blood pressures were similar between active and inactive recovery. The time courses of

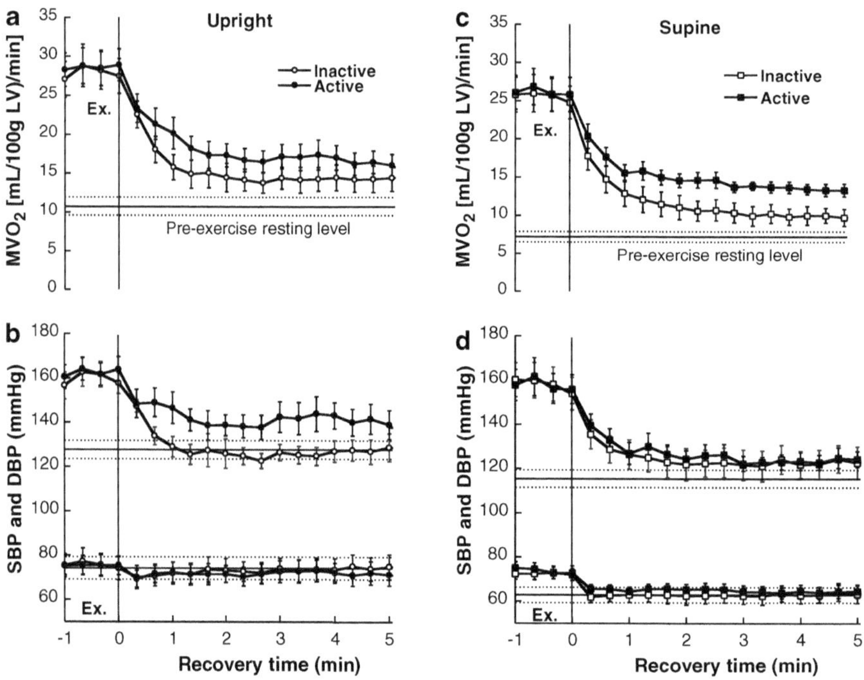

Fig. 5.4 Myocardial oxygen consumption (MVO_2, **a** and **c**) and systolic and diastolic blood pressures (SBP and DBP, **b** and **d**) during 10-min exercise at 60% of peak oxygen uptake and during inactive recovery and active recovery at 20% of peak oxygen uptake in upright position (left) and supine position (right). Data (mean ± SE) are consecutive 10-s averages obtained from seven subjects for each of the inactive- and active-recovery experiments in both positions. Vertical line (at $t = 0$) indicates the start of recovery. Horizontal thin and dotted lines represent pre-exercise resting levels as mean and ±SE, respectively

diastolic blood pressure were almost the same between active and inactive recovery in each of the upright and supine positions.

The difference between systolic blood pressure and diastolic blood pressure, referred to as arterial pulse pressure, during upright active recovery was significantly ($P < 0.05$) greater by about 10–17 mmHg than that during upright inactive recovery. The supine pulse pressure was not significantly different between active and inactive recovery.

5.4 Discussion

5.4.1 *Similar Arterial Pressures During Inactive and Active Recovery*

The time course of postexercise cardiac output was an initial fast decrease followed by a much slower decrease during inactive and active recovery, regardless of body position; however, the cardiac output was significantly greater during active than during inactive recovery in each of the upright and supine positions. The large cardiac output in the supine position during active recovery was predominantly due to heart rate, compared with inactive recovery. In the upright position, in contrast, the large cardiac output during active recovery was predominantly due to stroke volume, compared with inactive recovery. Systemic vascular conductance during active recovery was evidently higher than that during inactive recovery in each body position. Importantly, the heart rate, stroke volume, cardiac output, and systemic vascular conductance changed differently during inactive and active recovery from exercise in each body position. As a consequence, the time courses of mean arterial blood pressure were consistently similar between active and inactive recovery in each position, thus suggesting that mean arterial blood pressures were chiefly regulated.

5.4.2 *Outlines of Cardiovascular Regulation*

Before we discuss the important roles of cardiac output, the product of heart rate and stroke volume, in the circulation, a consideration of the heart rate in response to exercise in different postures helps us understand the integrated and coordinated control of the circulation along with an outline of cardiovascular regulation mechanisms. Resting heart rate is mainly regulated through the parasympathetic nerve outflow to the heart. On the start of exercise from rest, an initial increase in heart rate is caused by withdrawal of vagal nerve activity. This vagal withdrawal results from decreases in the tonic control of cardiac vagal motoneurons located in both the nucleus ambiguus and the caudal ventrolateral medulla in the medullary

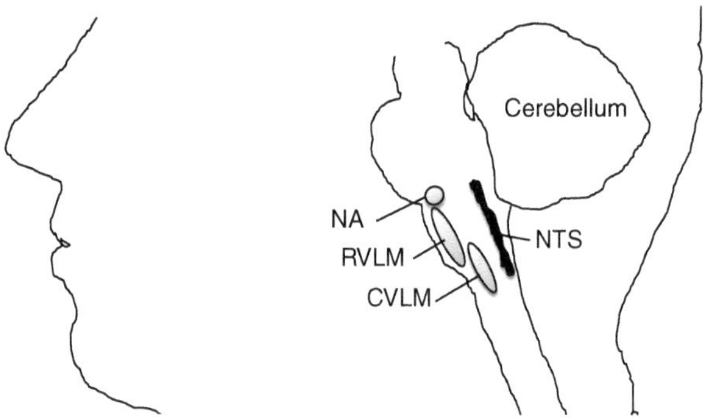

Fig. 5.5 Important areas involved in baroreflexes. *NA* nucleus ambiguus, *NTS* nucleus tractus solitarius, *RVLM* rostral ventrolateral medulla, *CVLM* caudal ventrolateral medulla

cardiovascular control center. These vagal motoneurons receive the inhibitory synaptic transmission from the nucleus tractus solitarius (NTS) that can be elicited by the centrally generated nerve signals, called central command, to contract working muscles, including locomotion (Fig. 5.5). The signals originating from subthalamic neurons activate the somatomotor and cardiovascular control systems separately at the onset of exercise. As the intensity of exercise increases, heart rate is further increased by sympathetic nerve activity switched from lower vagal tone. The degree of vagal nerve withdrawal would be proportional to perceived exertion or effort during exercise. When the autonomic nervous system is neutral, the heart rate is about 100 beats/min, at which the predominant innervation will transfer from the parasympathetic to the sympathetic nerves. Parenthetically, the heart has an intrinsic rhythmical depolarization of sinoatrial node cells at a rate of about 110 beats/min in the absence of any neurohumoral influence, although the rate tends to decline with age. From these pieces of physiological information, it is considered that the parasympathetic outflow to the heart affects the reduction of heart rate below 100 beats/min after 1 min of recovery period.

5.4.3 *Autonomic Nervous System During Light Exercise*

As mentioned above, when dynamic exercise is performed, heart rate usually increases by a certain amount, which is proportional to the exercise intensity or oxygen uptake. This linear increase in heart rate with exercise intensity, however, is not the case existing in Chap. 4. Briefly, when eight normal sedentary individuals in a sitting position performed cycle exercise at a light workload from rest, heart rate increased up to 86.3 ± 1.4 beats/min from 68.6 ± 1.2 beats/min at rest, and then gradually decreased to 72.1 ± 0.8 beats/min in 3 min of exercise. Since the contribution of cardiac sympathetic nerve activity usually makes a positive contribution to an increase in heart rate, not a decrease, the reduction in heart rate during constant light exercise is thought to be mainly due to reactivation of the vagal tone, suppression of the cardiac sympathetic nerve activity, or both. From those results, we have hypothesized that the recovery heart rate during light-intensity exercise in the upright position is modulated by a balance between the parasympathetic and sympathetic nerve activity via the reflex of cardiopulmonary mechanoreceptors. These receptors should be loaded by increased central venous volume with the assistance of the skeletal muscle pump.

In Chap. 4, power spectral analysis has revealed indirect evidence that the high-frequency (0.15–0.40 Hz) amplitude of heart rate variability, an index of vagal nerve activity, was significantly greater during light exercise than at rest in an upright sitting position. Also, the high-frequency amplitude during exercise was greater in the upright than the supine position, and vice versa at rest. At the time, supine heart rate increased in a stepwise manner from rest to exercise without a subsequently conspicuous decline. Interestingly, upright heart rate in the steady state is significantly lower than the supine heart rate during exercise. Inasmuch as the exercise is the same in the upright and supine positions, the neural inputs from central command to and Group III and IV afferent nerves of mechanical and chemical sensory receptors within the muscles to the cardiovascular control center in the medulla must be almost the same between the two postures. Since these neural inputs mediating the autonomic nervous system would be thus unlikely to induce such a difference in heart rate response to the same exercise, the other mechanism(s) should be considered in the response of heart rate to the exercise in comparison with the upright and supine positions. Interpretation of these problems will lead to clarification of the functions of the circulatory system during the recovery period in the following discussion.

5.4.4 Difference in Blood Flow Distribution

As for the different responses in heart rate between the upright and supine positions, it is necessary to consider a difference in the distribution of blood flow to the body under gravitational force. In a supine position at rest, the distribution of blood flow throughout the body depends on the relationship between blood vessel compliance and blood pressure. On the other hand, in a standing or sitting position at rest, the circulation is characterized by gravitational translocation of blood volume into dependent regions, in addition to the pressure-compliance relationship. Particularly, it has been pointed out that 500–700 mL of blood in the thoracic regions moves to the lower body in transition from the recumbent to the standing posture. This translocation of blood causes a decrease in arterial blood pressure due to a decrease in cardiac output. By feedback of unloaded arterial baroreceptor information, the heart rate and peripheral vascular resistance were reflexly increased to maintain homeostasis of blood pressure. In other words, during standing or sitting posture at rest, the central blood volume cannot be necessarily maintained to be the same as that in the supine position. The decrease in central venous pressure in the sitting and standing positions results in reductions in cardiopulmonary baroreflex signals to the cardiovascular control center, namely the nucleus tractus soritarius. The neural reflex signals from several mechanical receptors that detect tension present in the vascular wall of the large veins, the pulmonary artery and vein, or in the atrial and ventricular myocardium of the heart, collectively called cardiopulmonary mechanoreceptors, via small vagal nerves.

5.4.5 Reflex Mechanisms Included in the Central Nervous System

At rest in upright posture, a decrease in the tonic afferent of cardiopulmonary mechanoreceptors usually leads to augmentation in peripheral vascular sympathetic nerve tone to save arterial blood pressure from falling. The decreased firing rate of signal from the cardiopulmonary mechanoreceptors is of importance as a component of the reflex arc. This pathway signal relays from the vagal afferent nerves through the vasomotor center to the sympathetic adrenergic fibers terminating on the smooth muscle cells of the peripheral vasculature. In a human resting state, the sympathetic efferent activity originates from sympathetic premotor neurons in the rostral ventrolateral medulla (RVLM). The sympathetic neurons are tonically inhibited by the input of γ-aminobutyric acid (GABA) neurons from the caudal ventrolateral medulla (CVLM). Thus, the signals of stimulated cardiopulmonary mechanoreceptors and arterial baroreceptors would elicit the nucleus tractus soritarius neurons. These neurons in turn activate both cardiac vagal motoneurons in the nucleus ambiguus and the caudal ventrolateral medulla. The caudal ventrolateral medulla releases γ-aminobutyric acid to suppress the activity of the sympathetic premotor neurons in the rostral ventrolateral medulla (Fig. 5.5). It should be noted that there are also experimental reports that the increase or decrease of the tonic input from the cardiopulmonary receptor cells does not crucially affect either the increase or decrease in heart rate under the control of arterial baroreflexes. However, it has also been pointed out that the lower-pressure mechanosensory receptor input related to postural changes would modulate the reflex control of heart rate through other centrally neural afferents. To be more specific, the carotid sinus baroreflex may be more sensitive when cardiopulmonary receptors are unloaded by low central venous pressure. However, the reflex mechanisms of direct or indirect cardiopulmonary afferents that are involved in changes in heart rate are still somewhat controversial.

On the other hand, it has been argued that the two neural signals of the aortic arch and carotid sinus baroreceptors affect the balance of the sympathetic and parasympathetic branches of the autonomic nervous system. The two baroreceptor nerve signals are centrally sending information about the inclination of the body with the differences in the intensity and frequency of their neural discharges.

5.4.6 *Frank-Starling Mechanism*

On the postural change from supine to upright, a certain amount of central blood volume is shifted to the lower limbs under the effect of gravity, as mentioned before. As dynamic exercise of the lower limbs is performed, a large blood volume out of the venous system in the legs is emptied by the muscular contraction. In a case of upright exercise, thus, the return of blood enhanced by the muscle pump to the central venous system must allow the blood to fill the ventricles adequately by the Frank-Starling mechanism. The effect of muscle pump on the peripheral circulation simply appears as an increase in systemic vascular conductance. As shown in the middle panel of Fig. 5.3, the systemic vascular conductance during active recovery was higher than that during inactive recovery in both positions. The increased blood ejected from the heart (or increased stroke volume) and the decreased systemic vascular conductance in the upright position can lead to an increase in arterial pulse pressure (Fig. 5.4b). The increased pulse pressure effectively stimulates arterial baroreceptors, since they are more sensitive to dynamically fluctuating blood pressure than to a slow shift of mean arterial pressure.

5.4.7 *Baroreflex Pathways to Peripheral Vasculature*

The cardiopulmonary and arterial baroreceptors through the reflex pathways possibly elicit the vagal nerve tone and inhibit the sympathetic nerve tone to the effectors such as the heart and peripheral vasculature. The arterial baroreceptor afferents project to and synapse within the nucleus tractus soritarius. The activation of arterial baroreceptor afferents evokes the local corelease of glutamate and substance P that activate the nucleus tractus soritarius output neurons. The neural output excites cardiac vagal motoneurons in the nucleus ambiguus and the caudal ventrolateral medulla. Inhibitory γ-aminobutyric acid neurons in the excited caudal ventrolateral medulla also inhibit sympathetic premotor neurons in the rostal ventrolateral medulla (Fig. 5.5). In this manner, the sympathetic nerve efferents are totally under the control of arterial baroreflexes. A study of cardiovascular responses to lower body suction has demonstrated that reductions in arterial baroreceptor afferents initiate the rise in heart rate and the splanchnic vasoconstriction, whereas forearm (skin and muscle) vasoconstriction is elicited mainly by the cardiopulmonary baroreflexes.

5.4.8 Vagal Control Mechanism of Heart Rate

The withdrawal of vagal tone that accompanies voluntary exercise leads to increases in heart rate. Continuous limb movement is usually performed during active recovery with a light workload. Even if the exercise intensity is as light as possible, active recovery would be expected continuously to suppress somewhat cardiac vagal efferents, thereby maintaining a higher heart rate above the resting level. This assertion is in agreement with the results of this study on the supine active-recovery. In contrast, there was little difference in the time course of heart rate during active and inactive recovery in the upright position, the results of which are also in agreement with those of our previous studies. Chapter 2 has revealed that a vagal control mechanism plays an important role in reducing heart rate during the early and subsequent phases of recovery. Furthermore, prompt reduction in heart rate at rest after active-recovery exercise in the upright position can be mediated by vagal efferent nerve activity to the heart, as shown in Chap. 3. The cardiac vagal nerve traffic should be usually depressed by neural feedback from somatic afferents stimulated by muscle contraction. At the same time, contradictorily, the vagal efferent is activated by cardiopulmonary mechanoreceptor afferents induced by venous return, enhanced by the skeletal muscle pump. Also, the upright stroke volume is persistently elevated during active recovery from exercise by the Frank-Starling mechanism that is operative via the sustained venous return of blood to the heart by the muscle pump. Restoration of postexercise vagal nerve activity would thus be more enhanced under the background of autonomic stimulation during active recovery. This physiologic background is why heart rate promptly decreased by about 8 beats/min after the end of active-recovery exercise; however, such a fall in heart rate did not occur during inactive recovery, as shown in Fig. 1.5 of Chap. 1. It remains still unclear that the mechanisms of autonomic control by which the decreases in heart rate occur in a similar time course during active and inactive recovery in the upright position. Further studies are necessary to determine the contribution and balance of parasympathetic and sympathetic divisions of the autonomic nervous system to the changes in heart rate during recovery from exercise.

5.4.9 Effects of Venous Return on Stroke Volume

Upright stroke volume was persistently elevated during active recovery, but not during inactive recovery, the results of which are consistent with those of our and others' previous reports. As mentioned above, these results from the measurements suggest that the higher upright stroke volume during active recovery is due to augmentation of venous return induced by the skeletal muscle pump against the force of gravity, whereas the lower upright stroke volume during inactive recovery is due to pooling of venous blood in the legs. In contrast, supine stroke volume was almost unchangeable in the transitions from rest through exercise to recovery, likely because the skeletal muscle pump cannot operate effectively as central venous pressure and venous return are both already elevated. Importantly, the results of this study reveal that there is only a little difference in stroke volume between inactive and active recovery in supine posture.

5.4.10 Sympathetic Nerve Activity

It has been reported that in a supine position, the leg sympathetic nerve activity recorded by sympathetic microneurography remains unchangeable from rest through leg exercise to recovery. In contrast, the leg sympathetic activity during light-intensity exercise in an upright position significantly decreases below the levels of rest and inactive recovery. Importantly, the sympathetic nerve activity during upright exercise at light intensity is decreased to a level similar to that during supine exercise. The upright dynamic leg exercise increases cardiac filling pressure and then stimulates the cardiopulmonary mechanoreceptors, thereby resulting in suppression of the muscle sympathetic nerve activity. Accordingly, the difference in upright systemic vascular conductance during active and inactive recovery in the present study may be associated with loading and unloading, respectively, of cardiopulmonary mechanoreceptors. In contrast, the difference in supine systemic vascular conductance during active and inactive recovery is difficult to be explained as the contribution of muscle sympathetic activity with little difference in cardiopulmonary nerve afferents, since there was no significant difference in stroke volume. Thus, it may also be necessary to consider a reflex arising from the activation of chemosensory afferent nerve endings within the postexercise muscle, with the accumulation of venous blood. This reflex can more or less modulate the sympathetic vasoconstrictor nerve activity.

5.4.11 Contribution of Muscle Pump to the Circulation

Here we shall assess the function of the skeletal muscle, which is to pump blood from the arterial and venous circulatory systems in the exercise muscle using a simple model (Fig. 5.6). Movement of blood fluid can be assessed by calculating the pressure difference that drives the blood through a series of blood vessels with variable resistances. Blood flow can thus be described by mathematical expressions analogous to Ohm's law for electrical circuits. This model computes the total work (mean venous pressure times venous perfusion) of the muscle pump that helps to return the venous blood from the legs to the heart (see Appendix). To simplify the computation, it is assumed that the total peripheral vascular resistances are the same under the active- and inactive-recovery conditions; however, the substantial difference in pressure drop due to vasodilation during active recovery is determined to be a change in intravascular blood pressure (MBP_m) by muscle contraction. In the

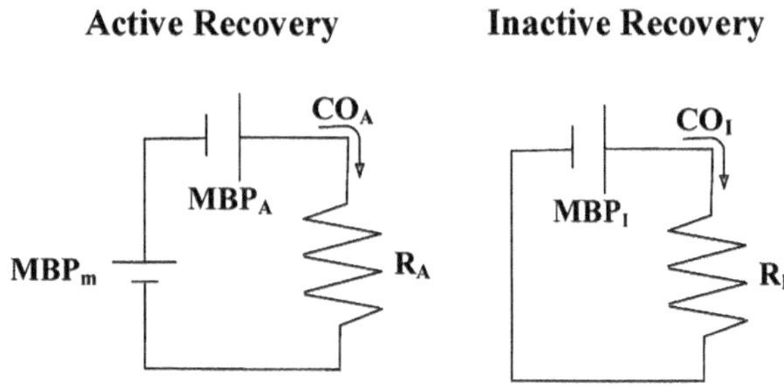

Fig. 5.6 A model of muscle pump work is analogous to Ohm's law for electric circuits (see Appendix for details). Adapted from Takahashi et al., reproduced by permission

upright position, the calculated value of driving pressure for venous perfusion produced by the muscle pump is 29.8 mmHg. If the central venous pressure is assumed to be 0 mmHg, the venous pressure, which becomes equivalent to the driving pressure in the foot, is comparable to other direct measurements of mean venous pressure in a range of 15–30 mmHg in the leg or foot. The mechanical power of the muscle pump (8.3 L/min × 29.8 mmHg) is estimated as 0.548 W (or 0.548 N ms^{-1}). Similarly, in the supine position, the driving pressure and mechanical power of the muscle pump are estimated as 13.2 mmHg and 0.300 W, respectively. The estimated value of venous pressure for supine active recovery is also similar to a mean venous pressure of 12–15 mmHg in the foot at rest in a supine man, reported previously in studies by other investigators. These results suggest that the muscle pump is more efficient in returning the venous blood to the heart in the upright than in the supine position.

5.4.12 Postexercise Hypotension

Numerous studies have demonstrated that a bout of exercise at mild to moderate intensity can lead to a postexercise decrease in arterial blood pressure in both upright and supine positions, which the phenomena is termed postexercise hypotension. The postexercise hypotension should be regulated by optimal alterations in parameters (e.g., cardiac output, heart rate, stroke volume, and systemic vascular conductance). These parameters are required for the maintenance of arterial blood pressure to transport nutrients, waste, and excess heat effectively through blood flow, regardless of body position, as shown in the previous chapters. In this study, the cardiovascular responses to the different recovery modes and body positions provide important information as to how functional changes in the variables are to regulate postexercise blood flow and pressure via reflexes of arterial and muscular sensory nerve afferents. Accordingly, the levels of postexercise blood pressure similar to the preexercise resting levels were observed during active and inactive recovery from exercise in each of the two positions. The cardiac output during active recovery from exercise also decreased more gradually than that during inactive recovery in each position. Thus, it seems to be important that the overall contribution of systemic vascular conductance to the regulation of postexercise blood pressure is greater in the active than in the inactive recovery, as shown in Fig. 5.3. The evidence for the role of higher systemic vascular conductance during active recovery in the control of postexercise blood pressure is that there exists a difference in the systemic vascular conductance between active and inactive recovery in the supine position, although the muscle pump is not efficient enough.

Although postexercise hypotension generally occurs in both normotensive and hypertensive individuals, it is more prominent in hypertensive individuals. In humans, postexercise hypotension has been observed in a variety of such activities as walking, cycling, running, swimming, and resistance exercise. This study examined the recovery of arterial blood pressure to the pre-exercise resting levels after a 10-min bout of moderate exercise in normotensive subjects. The exercise may not effectively produce postexercise hypotension. The reason for this result may be that the length of exercise period was too short to be effective. More prolonged exercise, for example, more than 20 min, has been documented to result in a greater decrease in blood pressure or postexercise hypotension for a longer period of time or for several hours.

5.4.13 Internalization of Neurokinin-1 Receptors

During exercise, Group III and IV nerve fibers or muscle mechanosensitive and chemosensitive afferent nerves terminate in the nucleus tractus solitaries. The somatic afferent signaling releases substance P in some interneurons. The action of substance P would be localized to excite the γ-aminobutyric acid inhibitory interneurons. The exited inhibitory interneurons lead to the suppression of the output of excitatory neurotransmission, glutamate, from the nucleus tractus solitarius into the nucleus ambiguus and the caudal ventrolateral medulla. This neuronal process resets arterial blood pressure to a higher level (exercise pressor response). In addition, binding of the substance P to neurokinin-1 receptors during exercise would trigger the receptors to undergo internalization. This internalization of the receptors weakens γ-aminobutyric acid interneuron excitability in the second-order barosensitive neurons of the nucleus tractus solitaries. This weakened signal resets a set point of baroreflexes to a lower level after cessation of exercise (postexercise hypotenison). After exercise, thus, reduced γ-aminobutyric acid inhibition in the nucleus tractus solitarius in the baroreflex pathway would signal the caudal ventrolateral medulla. Then the input of γ-aminobutyric acid receptors in the rostral ventrolateral medulla can decrease the baseline activity of the rostral ventrolateral neurons, thereby reducing sympathetic nerve activity and arterial blood pressure.

It is also important that the baroreceptor afferent fibers to the central nervous system play a crucial role in postexercise hypotension. The elimination of the cardiopulmonary and arterial baroreceptor nerves to the central nervous system in advance before the start of exercise attenuates the development of postexercise hypotension. Therefore, these baroreceptor inputs to the nucleus tractus solitarius can be the important components of the mechanism for exercise-induced internalization of the neurokinin-1 receptors located on the dendrites.

5.4.14 Risks of Heart Failure in Recovery from Exercise

The stroke volume during active recovery was greater in the supine than in the upright position. The greater stroke volume can be explained by increased preload for the diastole associated with increased central venous pressure and ventricular blood filling. It is also associated with decreased afterload for the systole through increases in systemic vascular conductance. In this manner, larger central venous volume in supine posture may overload the right and left atria and ventricles of the heart with blood. Such a crisis situation in dysfunctioning hearts, if any, may manifest as edema, especially in cardiac or pulmonary failure patients. Alternatively, the result that the heart rate during active recovery was higher than that during inactive recovery in the supine position was probably explained by the difference in vagal tone. The reduction in vagal nerve activity may lead to a risk of sudden cardiac arrest in patients with heart disease. Therefore, excessive filling of the postexercise atria and ventricles can be a risk factor for cardiac events in elderly adults with a decrease in myocardial distensibility and in heart failure patients with constrictive pericarditis effusion.

5.4.15 Myocardial Oxygen Consumption

Exercise is a typical stimulus for increased myocardial oxygen demand. During exercise, the main determinants of myocardial oxygen demand of the heart are as follows: heart rate, myocardial contractility, and ventricular work. An increase in oxygen demand of the left ventricle during exercise is met mainly by augmenting coronary blood flow and oxygen extraction. The heart maintains a very large extraction of oxygen. That is, 70–80% of the arterially delivered oxygen is extracted even at rest, as compared with 30–40% in skeletal muscle. Understandably, oxygen consumption by the heart is mainly required for contraction, with the maintenance of basal metabolism comprising only 10–20% of total myocardial oxygen consumption. The myocardial oxygen extraction would also reach 90% or more with a coronary venous oxygen tension of 15 mmHg during heavy exercise. The systemic and myocardial oxygen consumptions do not always change in parallel during a variety of exercise, dynamic and static exercise, for example.

The brachial arterial systolic blood pressure is usually higher than the central aortic blood pressure due to summation of the incident with the reflected pulse waves. Nevertheless, the estimation of myocardial oxygen consumption from the indirect brachial arterial systolic blood pressure has been reported to correlate well to that from the direct measurement of central aortic blood pressure at rest and during exercise. Despite these error factors, the noninvasive measurement with a pneumatic finger cuff should be substituted in normal individuals for an intra-arterial catheter without substantial loss in the ability to predict myocardial oxygen uptake. Heart rate and systolic arterial blood pressure can vary in different patterns depending on body posture and on the presence or absence of recovery exercise. Changes in myocardial oxygen uptake also qualitatively correlate well with those in cardiac output in this study. Collectively, the estimates of myocardial oxygen uptake by the noninvasive method for measuring heart rate, arterial pressure, and cardiac output can provide useful indicators for important information about myocardial blood flow, and cardiomyocyte metabolism as well.

5.4.16 Limitations

There may be a mechanism responsible for reflex control of venous tone during postural change. For example, extreme changes in arterial pressure, as would occur in massive hemorrhage, syncope, and cardiogenic shock, or in individuals subjected to thermal stress, cause venous tone to change reflexly. In fact, in the absence of considerable reductions in arterial pressure, no significant venoconstriction of the limbs is induced by baroreceptor-mediated reflexes while a person experiences short-term orthostatic stress. Therefore, the contribution of reflex venoconstriction to blood volume in the lower extremities may be insignificant in this study.

This study demonstrated that there was no significant difference in stroke volume between inactive and active recovery in the supine position. Interestingly, the supine stroke volume during inactive recovery was somewhat, but not significantly, greater than that during active recovery at the fifth minute; however, three out of seven subjects had lower stroke volume for inactive recovery. It is still undecided whether these differences in stroke volume have functional characteristics of the circulatory system during the recovery period.

The impedance cardiography, measuring stroke volume, used in this study is a good noninvasive technique, although it has some limitations regarding the measurement of absolute values. However, we have demonstrated in our previous study that impedance cardiography is suitable for application during posture change in the recovery period from exercise (see Chap. 2). The validity of impedance cardiography can be seen in the fact that the time courses of relative change in stroke volume during recovery from exercise in the upright and supine positions are consistent with those obtained with pulsed Doppler ultrasound. In this study, we achieved the reproducibility of estimates of stroke volume in individuals by taking care to attach the impedance electrodes to the same positions every time. To further improve both the day-to-day and intraindividual reproducibility, all experiments were repeated twice and an ensemble-average technique was adopted to the two measurements. These procedures are considered adequate to achieve our aim, although we are aware of a shortcoming in that the technique does not register beat-by-beat stroke volume. Therefore, further research on this issue is needed to enhance understanding of the responses of stroke volume and cardiac output to the interaction between posture change and limb movement during recovery from exercise by means of other more reliable techniques.

5.5 Summary

The effects of muscular contractions (the muscle pump) and body posture (the gravitational force) on cardiovascular responses were examined during recovery from moderate exercise in upright (sitting) and supine (lying) positions. Heart rate, stroke volume, cardiac output, arterial blood pressure, and systemic vascular conductance were measured in seven young male subjects at rest, during a 10-min bout of cycle exercise at 60% of peak oxygen uptake, and then during either complete rest for 5 min (inactive recovery) or cycling at 20% of peak oxygen uptake for 5 min (active recovery) in each of the upright and supine positions. In the upright position, heart rate had an initial rapid decrease and then a gradual decrease, the time courses of which were similar between inactive and active recovery. In contrast, in the supine position, heart rate during active recovery was significantly higher than that during inactive recovery. Upright stroke volume during inactive recovery decreased gradually to the pre-exercise resting level, whereas that during active recovery remained significantly elevated. Supine stroke volumes that were similar between inactive and active recovery were not significantly different from the pre-exercise resting levels. Thus, the higher stroke volume in the upright position and the higher heart rate in the supine position both resulted in significantly greater cardiac outputs during active than during inactive recovery. In addition, systemic vascular conductance was greater during active than during inactive recovery, regardless of body position. As a result, mean blood pressures were almost the same between active and inactive recovery in each position. This is because mean blood pressure is the product of cardiac output and reciprocal systemic vascular conductance, which means, the two variables compensate for each other to maintain similar mean blood pressures during active and inactive recovery. Despite extra continuous limb movements, upright heart rate during active recovery with larger stroke volume was similar to that during inactive recovery with smaller stroke volume. In contrast, supine stroke volume during active recovery with higher heart rate was similar to that during inactive recovery with lower heart rate. The higher stroke volume during active recovery in the upright position would be due to the venous return, enhanced by the muscle pump, but that was not the case in the supine position. These results reveal that the venous return to the heart due to the muscle pump is very important in cardiac vagal tone via the cardiopulmonary mechanoreflex during active recovery in the upright position. However, the effect of the muscle pump is hardly exhibited in the supine position.

Appendix

As shown in Fig. 5.6, we employ a model for simulating the effect of the muscle pump on the systemic blood circulation. This simple simulation model is analogous to Ohm's law for electric circuits, described below. Let *CO*, *MBP*, and *R* denote cardiac output, mean blood pressure, and total peripheral resistance, respectively. Subscripts "A" and "I" refer to measurements during active recovery and inactive recovery, respectively. Driving mean blood pressure (MBP_m) is the pressure generated by muscular contractions. The total peripheral vascular resistance is the algebraic sum of the artery, capillary, and vein resistances. The resistances R_A and R_I are supposed to be the same and are then represented by '*R*.' An *MBP* of 94 mmHg is the average of MBP_A and MBP_I (Table 5.1). Ohm's law leads analogically to the following equations:

$$\mathrm{CO}_A \cdot R = \mathrm{MBP} + \mathrm{MBP}_m$$

and

$$CO_I \cdot R = MBP,$$

where the mean values of 8.3 L·min^{-1} and 6.3 L·min^{-1} measured in this study are entered into CO_A and CO_I, respectively, in the upright position. The equations are then rearranged by eliminating *R* as,

$$\mathrm{MBP}_m = \frac{(\mathrm{CO}_A - \mathrm{CO}_I)}{\mathrm{CO}_I} \times \mathrm{MBP} = 29.8\,\mathrm{mmHg}.$$

Mechanical power of the muscle pump is expressed as *P*:

$$P = \mathrm{CO}_A \cdot \mathrm{MBP}_m = 0.548\,\mathrm{W}.$$

Similarly, for the supine position, the pressure produced by the muscle pump and the mechanical power of the muscle pump are as follows:

$$\mathrm{MBP}_m = \frac{(10.3 - 8.9)}{8.9} \times 84 = 13.2\,\mathrm{mmHg}$$

and

$$P = \mathrm{CO}_A \times \mathrm{MBP}_m = 0.300\,\mathrm{W}.$$

Further Reading

Takahashi T, Hayano J, Okada A, Saitoh T, Kamiya A. Effects of the muscle pump and body posture on cardiovascular responses during recovery from cycle exercise. Eur J Appl Physiol. 2005;94:576–83.

Asmussen E, Nielsen M. Cardiac output during muscular work and its regulation. Physiol Rev. 1955;35:778–800.

Bevegård S, Castenfors J, Lindblad LE, Tranesjö J. Blood pressure and heart rate regulating capacity of the carotid sinus during changes in blood volume distribution in man. Acta Physiol Scand. 1977;99:300–12.

Bjurstedt H, Rostenhamer G, Balldin U, Katkov V. Orthostatic reactions during recovery from exhaustive exercise of short duration. Acta Physiol Scand. 1983;119:25–31.

Boer P, Roos JC, Geyskes GG, Dorhout Mees EJ. Measurement of cardiac output by impedance cardiography under various conditions. Am J Phys. 1979;237:H491–6.

Boscan P, Kasparov S, Paton JFR. Somatic nociception activates NK_1 receptors in the nucleus tractus solitarii to attenuate the baroreceptor cardiac reflex. Eur J Neurosci. 2002;16:907–20.

Brown SP, Li HE, Chitwood LF, Anderson ER, Boatwright D. Blood pressure, hemodynamic, and thermal responses after cycling exercise. J Appl Physiol. 1993;75:240–5.

Carter R, Watenpaugh DE, Smith ML. Genome and hormones: gender differences in physiology selected contribution: gender differences in cardiovascular regulation during recovery from exercise. J Appl Physiol. 2001;91:1902–7.

Carter R, Watenpaugh DE, Wasmund WL, WasmundSL SML. Muscle pump and central command during recovery from exercise in humans. J Appl Physiol. 1999;87:1463–9.

Chandler MP, DiCarlo SE. Sinoaortic denervation prevents postexercise reduction in arterial pressure and cardiac sympathetic tonus. Am J Phys. 1997;273:H2734–45.

Chen C-Y, Bechtold AG, Tabor J, Bonham AC. Exercise reduces GABA synaptic input onto nucleus tractus solitarii baroreceptor second-order neurons via NK1 receptor internalization in spontaneously hypertensive rats. J Neurosci. 2009;29:2754–61.

Chen C-Y, Bonham AC. Postexercise hypotension: central mechanisms. Exerc Sport Sci Rev. 2010;38:122–7.

Cléroux J, Kouamé N, Nadeau A, Coulombe D, Lacourcière Y. Aftereffects of exercise on regional and systemic hemodynamics in hypertension. Hypertension. 1992;19:183–91.

Collins HL, DiCarlo SE. Attenuation of postexertional hypotension by cardiac afferent blockade. Am J Phys. 1993;265:H1179–83.

Degtyarenko AM, Kaufman MP. Barosensory cells in the nucleus tractus solitarius receive convergent input from group III muscle afferents and central command. Neuroscience. 2006;140:1041–50.

Duncker DJ, Bache RJ. Regulation of coronary blood flow during exercise. Physiol Rev. 2008;88:1009–86.

Epstein SE, Beiser GD, Stampfer M, Braunwald E. Role of the venous system in baroreceptor-mediated reflexes in man. J Clin Invest. 1968;47:39–152.

Evert TJ, Smith JJ, Barney JA, Merrill DC, Smith GK. The use of thoracic impedance for determining thoracic blood volume changes in man. Aviat Space Environ Med. 1986;57:49–53.

Flamm SD, Taki J, Moore R, Lewis SF, Keech F, Maltais F, Ahmad M, Callahan R, Dragotakes S, Alpert N, Strauss HW. Redistribution of regional and organ blood volume and effect on cardiac function in relation to upright exercise intensity in healthy human subjects. Circulation. 1990;81:1550–9.

Folkow B, Neil E. Circulation. New York: Oxford University Press; 1971.

Freund PR, Rowell LB, Murrhy TM, Hobbs SF, Butler SH. Blockade of the pressor response to muscle ischemia by sensory nerve block in man. Am J Phys. 1979;237:H433–9.

Gauer OH, Thron HL. Postural changes in the circulation. In: Hamilton WF, Dow P, editors. Handbook of physiology, vol. 3. Washington: American Physiological Society; 1965. p. 2409–39.
Goodwin GM, McCloskey DI, Mitchell JH. Cardiovascular and respiratory responses to changes in central command during isometric exercise at constant muscle tension. J Physiol. 1972;226:173–90.
Harris P. Role of arterial pressure in the oedema of heart disease. Lancet. 1988;331:1036–8.
Heymans C, Neil E. Reflexogenic areas of the cardiovascular system. London: J & A Churchill Ltd.; 1958.
Imai K, Sato H, Hori M, Kusuoka H, Ozaki H, Yokoyama H, Takeda H, Inoue M, Kamada T. Vagally mediated heart rate recovery after exercise is accelerated in athletes but blunted in patients with chronic heart failure. J Am Coll Cardiol. 1994;24:1529–35.
Hellebrandt FA, Franssen EB. Physiological study of the vertical stance of man. Physiol Rev. 1943;23:200–55.
Higginbotham MB. Cardiac performance during submaximal and maximal exercise in healthy persons. Heart Fail. 1988;4:68–76.
Imholz BPM, Settls JJ, Van Meiracker AH, Wesseling KH, Wieling W. Non-invasive continuous finger blood pressure measurement during orthostatic stress compared to intra-arterial pressure. Cardiovasc Res. 1990;24:214–21.
Jacobsen TN, Morgan BJ, Scherrer U, Vissing SF, Lange RA, Johnson N, Steves Ring W, Rahko PS, Hanson P, Victor RG. Relative contributions of cardiopulmonary and sinoaortic baroreceptors in causing sympathetic activation in the human skeletal muscle circulation during orthostatic stress. Circ Res. 1993;73:367–78.
Johnson EC, Hudson TL, Greene ER. Left ventricular hemodynamics during exercise recovery. J Appl Physiol. 1990;69:104–11.
Kajekar R, Chen C-Y, Mutoh T, Bonham AC. $GABA_A$ receptor activation at medullary sympathetic neurons contributes to postexercise hypotension. Am J Phys. 2002;282:H1515–624.
Kenney MJ, Seals DR. Postexercise hypotension: key features, mechanisms, and clinical significance. Hypertension. 1993;22:653–64.
Kubicek WG, Karnegis JM, Patterson RP, Witsoe DA, Mattson RH. Development and evaluation of an impedance cardiac output system. Aerosp Med. 1966;37:1208–12.
Kulics JM, Collins HL, DiCarlo SE. Postexercise hypotension is mediated by reductions in sympathetic nerve activity. Am J Phys. 1999;276:H27–32.
Laughlin MH. Skeletal muscle blood flow capacity: role of muscle pump in exercise hyperemia. Am J Phys. 1987;253:H993–H1004.
Laughlin MH, Davis MJ, Secher NH, van Lieshout JJ, Arce-Esquivel AA, Simmons GH, Bender SB, Padilla J, Bache RJ, Merkus D, Duncker DJ. Peripheral circulation. Compr Physiol. 2012;2:321–447.
Loeppky JA, Greene ER, Hoekenga DE, Caprihan A, Luft UC. Beat-by-beat stroke volume assessment by pulsed Doppler in upright and supine exercise. J Appl Physiol. 1981;50:1173–82.
Mark AL, Mancia G. Cardiopulmonary baroreflexes in humans. In: Shepherd JT, Abboud FM, Geiger SR, editors. Handbook of physiology. The cardiovascular system. Peripheral circulation and organ blood flow, vol. 3. Bethesda: American Physiological Society; 1983. p. 795–813.
Miyamoto Y, Higuchi J, Abe Y, Hiura T, Nakazono Y, Mikami T. Dynamics of cardiac output and systolic time intervals in supine and upright exercise. J Appl Physiol. 1983;55:1674–81.
Miyamoto Y, Tamura T, Hiura T, Nakamura T, Higuchi J, Mikami T. The dynamic response of the cardiopulmonary parameters to passive head-up tilt. Jpn J Physiol. 1982;32:245–58.
Nelson RR, Gobel FL, Jorgensen CR, Wang K, Wang Y, Taylor HL. Hemodynamic predictors of myocardial oxygen consumption during static and dynamic exercise. Circulation. 1974;50:1179–89.
Paterson DJ. Antiarrhythmic mechanisms during exercise. J Appl Physiol. 1996;80:1853–62.

Persson PB. Modulation of cardiovascular control mechanisms and their interaction. Physiol Rev. 1996;76:193–244.
Pollack AA, Wood EH. Venous pressure in the saphenous vein at the ankle in man during exercise and changes in posture. J Appl Physiol. 1949;1:649–62.
Potts JT. Inhibition neurotransmission in the nucleus tractus solitarii: implications for baroreflex resetting during exercise. Exp Physiol. 2006;91:59–72.
Ray CA, Rea RF, Clary MP, Mark AL. Muscle sympathetic nerve responses to dynamic one-legged exercise: effect of body posture. Am J Phys. 1993;264:H1–7.
Raven PB, Fadel PJ, Ogoh S. Arterial baroreflex resetting during exercise: a current perspective. Exp Physiol. 2005;91:37–49.
Rothe CF. Venous system: physiology of the capacitance vessels. In: Shepherd JT, Abboud FM, Geiger SR, editors. Handbook of physiology. The cardiovascular system: peripheral circulation and organ blood flow, vol. 3. Bethesda: American Physiological Society; 1983. p. 397–452.
Rowell LB. Human cardiovascular control. New York: Oxford University Press; 1993.
Rowell LB, O'Leary DS. Reflex control of the circulation during exercise: chemoreflexes and mechanoreflexes. J Appl Physiol. 1990;69:407–18.
Saitoh T, Nakajima T, Takahashi T, Kawahara K. Changes in cardiovascular function on inhibition of apoptosis in left ventricular remodeling after myocardial infarction. Cardiovasc Pathol. 2006;15:130–8.
Siebenmann C, Rasmussen P, Sørensen H, Zaar M, Hvidtfeldt M, Pichon A, Secher NH, Lundby C. Cardiac output during exercise: a comparison of four methods. Scand J Med Sci Sports. 2015;25:e20–7.
Slinker BK, Glantz SA. End-systolic and end-diastolic ventricular interaction. Am J Phys. 1986;251:H1062–75.
Smith JJ, Bush JE, Wiedmeier VT, Tristani FE. Application of impedance cardiography to study of postural stress. J Appl Physiol. 1970;29:133–7.
Starling EH. The fluid of the body. London: Archibald Constable; 1909.
Stegall HF. Muscle pumping in the dependent leg. Circ Res. 1966;19:180–90.
Takahashi T. Easy to understand data analysis and statistics: let's start analysis statistics of medical and life science with excel. Tokyo: Ohmsha Ltd; 2017.
Takahashi T, Ashikawa K, Okada A, Takeshima N. Hypothesis: a role of cardiac receptor nerve afferent in reflex control of heart rate during light exercise in upright humans. Ther Res. 2003;24:1583–93.
Takahashi T, Miyamoto Y. Influence of light physical activity on cardiac responses during recovery from exercise in humans. Eur J Appl Physiol. 1998;77:305–11.
Takahashi T, Okada A, Hayano J, Takeshima N. Responses of heart rate and vagal tone to treadmill walking on land and in water in healthy elderly people. JAPA. 2003;11:18–26.
Takahashi T, Okada A, Hayano J, Tamura T. Influence of cool-down exercise on autonomic control of heart rate during recovery from dynamic exercise. Front Med Biol Eng. 2002;11:249–59.
Takahashi T, Okada A, Hayano J, Tamura T, Miyamoto Y. Influence of duration of cool-down exercise on recovery of heart rate in humans. Ther Res. 2000;21:170–5.
Takahashi T, Okada A, Saitoh T, Hayano J, Miyamoto Y. Difference in human cardiovascular response between upright and supine recovery from upright cycle exercise. Eur J Appl Physiol. 2000;81:233–9.
Takahashi T, Shibata M, Kamiya A. Mechanism of macromolecule concentration in collecting lymphatics in rat mesentery. Microvasc Res. 1997;54:193–205.
Takahashi T, Yanagida H, Wang S, Matsuo T, Okada A, Kuwayama T, Natsui J. Minimum oxygen cost of human walking with geometrically similar leg movements. Ther Res. 2009;30:113–24.
Terziotti P, Schena F, Gulli G, Cevese A. Post-exercise recovery of autonomic cardiovascular control: a study by spectrum and cross-spectrum analysis in humans. Eur J Appl Physiol. 2001;84:187–94.

Victor RG, Leimbach WN. Effects of lower body negative pressure on sympathetic discharge to leg muscles in humans. J Appl Physiol. 1987;63:2558–62.
Volianitis S, Yoshga CC, Vogelsang T, Secher NH. Arterial blood pressure and carotid baroreflex function during arm and combined arm and leg exercise. Acta Physiol Scand. 2004;181:289–95.
Wang Y, Marshall RJ, Shepherd JT. The effect of changes in posture and of graded exercise on stroke volume in man. J Clin Invest. 1960;39:1051–61.
Williamson JW. The relevance of central command for the neural cardiovascular control of exercise. Exp Physiol. 2010;95:1043–8.

Chapter 6
Blood Flow Velocity in Femoral Artery

6.1 Introduction

We have so far extensively studied that the cardiovascular regulation during recovery from exercise is mediated by a variety of factors, including the autonomic nervous system, baroreceptor reflexes, neural afferents originating in muscular mechanoreceptors and chemoreceptors (or metaboreceptors), and central command from the cerebral motor cortex. Especially, in an upright position, the functions of the skeletal muscle pump play important roles in maintaining appropriate blood flow to postexercise muscles and in promoting the venous return of blood to the heart. These functions thus enhance cardiac filling and augment stroke volume via the Frank-Starling mechanism. However, understanding of postexercise hemodynamics, such as the rate of blood flow to exercising muscles in humans, has been still limited. The reason for this limitation is that it is difficult for us to measure the velocity of blood flow in conduit vessels through working muscles accurately and noninvasively; for example, measuring sensors may be displaced from target vessels owing to muscle contraction and relaxation.

A Doppler ultrasound technique can measure blood flow velocity noninvasively in the femoral artery during dynamic knee extensor and flexor exercise. Apart from such continuous muscular contraction in leg exercise, the technique has been often utilized to estimate limb blood flow intermittently between muscle contractions or just after the termination of exercise. Even though Doppler-ultrasound methods have offered resistance against motion artifacts, it may still be difficult to measure blood flow velocity continuously in the large arteries and veins during dynamic exercise of large muscle groups.

This chapter is concerned with continuous measurement of blood velocity in the femoral artery during dynamic exercise by use of a Doppler ultrasound method. A new experimental system that we have developed is applied to the continuous measurement in voluntary participants who perform the knee extension and flexion of a

T. Takahashi, *The Cardiovascular System During Exercise and Recovery*,
https://doi.org/10.1007/978-981-99-8486-2_6

two-legged dynamic exercise. The knee extension and flexion of both legs are transformed to movement of pedaling by means of a mechanical mechanism of the system, including a set of heel harnesses with aluminum rods connected to the pedals of a cycle ergometer. If the mechanism is used in reverse, the subject's legs can be passively moved as the extension and flexion movements through the link motion by experimenter's pedaling on the cycle ergometer. The knee extension and flexion in this way allows us to measure blood flow velocity continuously in the femoral artery without serious artifacts of Doppler-signal errors.

Thus, the present study is designed to assess the effects of active and passive leg movements on femoral artery blood flow during recovery from exercise of dynamic knee extension and flexion in an upright seated position, as compared with the effect of no leg movements during a quietly resting recovery. Even though the participants move their legs against very low resistance during a period of recovery from exercise, increased stroke volume above the resting level is maintained through venous return from lower limbs with the assistance of the muscle pump, as shown in the previous chapters. So, we hypothesize that femoral artery blood flow during postexercise recovery is increased by passive leg movement as well as active leg movement. In-depth knowledge of difference in muscle blood flow dynamics between passive and active recovery is fundamental for an understanding of the cardiovascular system to control blood pressure. Importantly, passive muscular contraction has also been used to isolate somatic neurogenic inputs effectively to the central nervous system, or the cardiovascular control center, in noninvasive experiments on humans.

The cardiovascular system can be in concert to control blood pressure and flow with various sports or physical activities in our daily living. The regulation of the circulatory system with respect to recovery does not always follow the regulation at the beginning of exercise simply in the opposite way. Moreover, relatively little is known about the differences in the effects of inactive, passive, and active recovery modes after exercise on blood velocity in the femoral artery. This chapter also addresses the clinically and scientifically important data on how muscle blood flow velocity is affected by the three different modes during recovery from exercise. Furthermore, information about hemodynamics, including wall shear stress to the lumen of vessels, provides insight into the salutary and vicious effects of postexercise hyperemia on the vasculature of the working muscles.

6.2 Materials and Methods

6.2.1 Human Subjects

Five healthy young male volunteers participated in this study. The physical characteristics of the subjects, averaging 22.2 ± 0.1 (mean ± SE) years old, were as follows: a mean height of 1.69 ± 0.02 m, a mean body mass of 59.2 ± 0.7 kg, and a mean peak oxygen uptake of 2.55 ± 0.11 L/min. All the participants were fully informed about the procedures, risks, and benefits of the study. They gave a written informed consent before the experiments. None of them had any history of circulatory or respiratory disease. The research and experiment was conducted with the approval of the Health Research Ethics Board at the university. It was also conducted in compliance with the principles outlined in the Declaration of Helsinki.

None of the subjects was engaged in any regular physical activity or involved in competitive sports. They were measured for their peak oxygen uptake to perform a progressive exercise on a cycle ergometer with an electromagnetic-braked system (Corival 300; Lode, The Netherlands) at an incremental work rate of 15 W/min until their exhaustion; for more details, see Chap. 1.

6.2.2 Measurements of Cardiorespiratory Variables

Oxygen uptake (VO_2, in liters per minute) was measured breath-by-breath with the use of an on-line computerized measurement system. A detailed description of measurement has been given in the previous chapters. Briefly, ventilatory airflow was monitored with a hot-wire-type pneumotachograph (FR-2; Minato, Osaka, Japan). The composition of expired gas was continuously analyzed with a medical mass spectrometer (WSMR-1400; Westron, Chiba, Japan). The mass spectrometer was calibrated with a standard reference gas mixture before each measurement.

Heart rate (*HR*, in beats per minute), stroke volume (*SV*, in milliliters), and cardiac output (*CO*, in liters per minute) were also measured with the on-line computerized measurement system. Cardiac output was recorded by impedance cardiography (ICG). An impedance cardiograph device (RGA-5; Nihon Koden, Tokyo, Japan) applied a constant, sinusoidal current of 350 μA to two outside electrodes of silver tape, 5 mm wide, and recorded the electrical impedance between the other two inside electrodes. One pair of parallel electrodes with 30-mm distance was placed on the neck and the other pair of parallel electrodes with 30-mm distance was placed on the torso.

Stroke volume was calculated from a Kubicek standard equation. Artifacts incidental to respiration and body motion on impedance signals were removed arithmetically by averaging sequential ICG waveforms over 10 cardiac cycles. Absolute values for stroke volume derived from impedance signal cannot be completely determined. However, it is important to recognize that ICG is well suitable for estimating relative changes in stroke volume from rest to exercise and recovery. In particular, we checked the reproducibility of estimates of stroke volume within each individual in response to exercise from day to day. Care was taken to locate the electrode positions for ICG at the same position, since the stroke volume measured is a function of the distance between the inner electrodes (see Eq. (1.1) in Chap. 1).

Arterial blood pressure (in millimeters of mercury, mmHg) was measured noninvasively with a pneumatic finger cuff (2300 Finapres blood pressure Monitor; Ohmeda, NEC San-ei Ltd., Tokyo, Japan). Mean arterial blood pressure (*MBP*) was calculated as diastolic arterial pressure plus one-third of the pulse pressure: $MBP = DBP + (SBP - DBP)/3$, where SBP and DBP are systolic blood pressure and diastolic blood pressure, respectively. The arm and hand that were supported with an arm rest of a chair for exercise tests were held at heart level to avoid variations of blood pressure due to changes in hydrostatic pressure.

Total peripheral vascular resistance (*TPR*, $mmHg \cdot L^{-1} \cdot min$) was calculated by dividing mean arterial blood pressure by cardiac output.

Blood flow velocity in the femoral artery was continuously measured with a pulsed Doppler ultrasound velocimeter (DVM-4000; Nihon-Koden Co., Tokyo, Japan). A flat transducer of piezoelectric ceramics with an operating frequency of 5 megahertz (MHz) was fixed to the skin over the superficial branch of the common femoral artery at about 5 cm away from the inguinal ligament. The position of the transducer was chosen to minimize interference in measurement by disturbance due to leg movement during exercise. The angle of the transducer to the skin was approximately 60°. An ultrasound gate was maintained at full width to facilitate insonication of the total width of the artery with transmitted signals of constant intensity. Optimal positioning of the Doppler device was verified by our observing the velocity spectra on a PC monitor, thereby resulting in clear Doppler signals during rest, exercise, and recovery.

The velocity signal was recorded at 500 Hz on a computer system along with an electrocardiograph to allow beat-by-beat analysis of the data. The mean beat-by-beat blood flow velocity (*MBV*) was calculated by integrating the total area under the profile of instantaneous blood flow velocity over the end of one heartbeat and the beginning of the next in the QRS wave of the electrocardiogram. The output signals of femoral artery blood flow velocity and R–R interval monitored with the Doppler equipment were sequentially translated to the personal computer through an analog-to-digital converter, RS232C interface; however, no volumetric rate of muscle blood flow was obtained, since this equipment was unable to measure the caliber of the femoral artery or the cross-sectional area of target vessels.

6.2.3 Ergometry System of Leg Exercise

Dynamic exercise with knee extensor and flexor muscles was carried out at a cadence rate of 60 rpm (revolutions per minute) with the use of the cycle ergometer. The layout of experiment is schematized in Fig. 6.1. The subjects sat in a chair in front of the cycle ergometer. The chair was set at an appropriate height to allow the subject's legs to move freely. The body trunk was vertical and both thighs were horizontal. The chair was adjusted so that both the legs could be in neutral vertical position; that is, the knee-joint angle between the thigh and leg was approximately 90°. The movement of knee extension and flexion of the legs was transformed to a mode of pedaling via a pair of heel harnesses with individual single aluminum rods connected to the pedals of the cycle. The harness and rod for each leg weighed a total of 2 kg. In practical usage, however, the weight loaded on the subject's leg could be

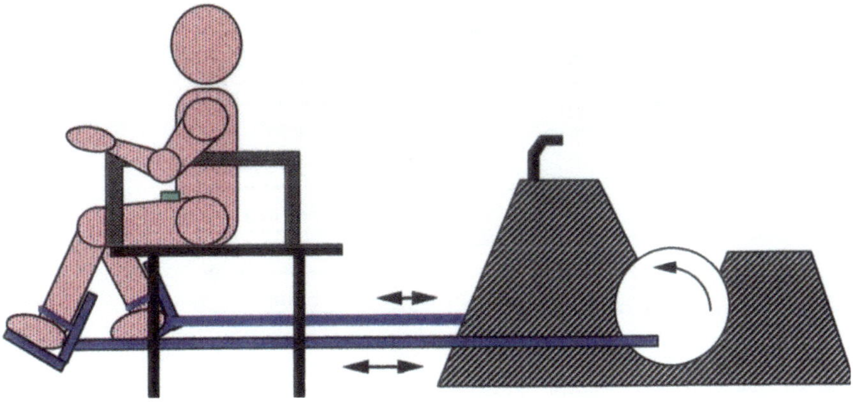

Fig. 6.1 Diagram of experimental layout with subject seated in a chair in front of a cycle ergometer. Adapted from Takahashi et al., reproduced by permission

reduced by about 1 kg, because the end of the rod was supported by one pedal of cycle ergometer. As for the forward and backward movements of the legs, the range of motion of the legs was within 30° across the vertical line, while the two legs moved inversely from each other.

The exercise that the subjects involuntarily performed was referred to as passive leg exercise or movement. The passive leg exercise was induced by an experimenter, who normally performed pedaling on the cycle ergometer. The degrees of knee joint motion during passive leg exercise were adjusted to be the same as those achieved during voluntary leg exercise. The knee extension and flexion was carried out smoothly and continuously in a manner analogous to stationary cycling. The frequency of leg movements was maintained at a rate of 60 rpm throughout exercise with the aid of an audible metronome. The pedaling rate of the bicycle was monitored with an inductive revolution meter fitted to the ergometer. Each participant was allowed to practice and become skilled at knee extension and flexion exercise on a separate day about 1 week before the formal study.

6.2.4 Experimental Protocols

The subjects rested for 2 min in an upright seated position on the chair in order to obtain baseline levels of cardiorespiratory variables (Fig. 6.2). They performed knee extension and flexion exercise at a work rate of 0 W during a warm-up period of 3 min and further at a work rate of 60 W for 5 min. The recovery phase during a 5-min period after exercise was examined for cardiorespiratory variables in response to three different recovery modes: (1) quietly sitting rest with no leg movement (inactive recovery); (2) passive-performed leg movements (passive recovery); and (3) voluntary-performed leg movements at 0 W (active recovery). The subjects were

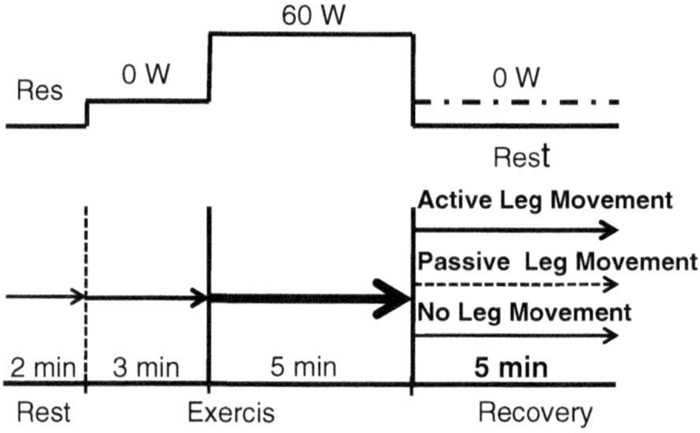

Fig. 6.2 Schematic representation of exercise protocols. Exercise was performed for 5 min at a work rate of 60 W after 2-min rest and 3-min exercise at 0 W. The subsequent recovery during a 5-min period was undergone in the following three modes: (1) no leg movement (inactive recovery), (2) involuntary leg movement (passive recovery), or (3) voluntary leg movement (active recovery) at 0 W

in the upright sitting position throughout the experiments. They were informed that the tests included one of the inactive, passive, and active modes during recovery after stimulus exercise; however, the order of tests was randomized. Exercise intensities of 0 W and 60 W were equivalent to relative intensities of 12 ± 2% and 31 ± 3%, respectively, of peak oxygen uptake. The changes in work rate were automatically controlled by a computer program.

Each exercise test was started at 10:00 for all subjects, who were randomly assigned to the three different modes of inactive, passive, and active recovery after the same exercise. They completed two repetitions of three exercise tests with the different modes of recovery; specifically, they performed one test a week to avoid continuous exercise tests, which have a training effect.

The subjects were instructed to abstain from any beverages containing caffeine or alcohol from 21:00 the night before exercise tests and to abstain from vigorous exercise for 36 h before the exercise tests. They finished a small breakfast before 7:00 on the test day.

All experiments were conducted at a room temperature of 21–22 °C.

6.2.5 Data Analysis

Lagrange polynomials were fitted to breath-by-breath and beat-by-beat data for cardiorespiratory variables, which were interpolated regularly at 5-s intervals. The two interpolated data for two exercise tests with the same mode of recovery were averaged to yield a single set of data for each subject. Group mean values were obtained from the five sets of individual data. Grand mean values for all variables were determined from the last minute of data at rest before exercise, during exercise, and during recovery.

6.2.6 Statistical Analysis

The Friedman repeated measures analysis of variance on rank and Wilcoxon's signed-rank (post hoc) test were used to assess the differences in mean values for variables at the fifth minute of recovery between three recovery modes. Cardiorespiratory responses in the postexercise transition during the first-minute period of recovery phase were examined by use of one-way repeated-measures analysis of variance. Differences in the magnitude of peak of mean blood velocity and in the magnitude of bottom of total peripheral vascular resistance, which appeared early during the time course of recovery, were assessed with Dunnett's multiple comparison test. A least-squares regression line was fitted to individual data points of mean blood flow velocity for the first run plotted against those for the second run together at rest and during exercise at 0 and 60 W. The measurement repeatability of mean blood velocity in the two runs was examined with Pearson's product-moment correlation coefficient. The differences were considered significant for all statistical analyses at $P < 0.05$. All values are expressed as mean ± standard error (SE).

6.3 Results

The cardiovascular variables that control the circulation during the recovery period were described in detail; the variables in response to rest and exercise were also shown in graphs for reference to those during recovery.

Table 6.1 summarizes the mean values for all variables obtained from the five subjects at rest, during mild exercise, and during inactive, passive, and active recovery. The workloads on exercise and active recovery were equivalent to relative intensities of about 31% and 12%, respectively, of peak oxygen uptake. There were essentially no significant differences in any mean values for the same rest or exercise between the three experiments.

Figure 6.3 shows the changes in oxygen uptake for the five subjects during the same rest and exercise, but with three different modes of inactive, passive, and active recovery. Oxygen uptakes for all of the three recoveries rapidly decreased during the early phase of recovery within the first 2 min. After rapid decreases, however, the oxygen uptakes slowly decreased. The mean oxygen uptake at the fifth minute was significantly ($P < 0.05$) greater during active than during inactive and passive recovery; however, there was no significant difference in oxygen uptake between inactive and passive recovery.

Figure 6.4a shows that recovery heart rate rapidly decreased without any delay after exercise, regardless of recovery mode. The subsequently decreased heart rate during active recovery reached a steady-state level at the warm-up exercise of 0-W workload, whereas those during inactive and passive recovery reached pre-exercise resting levels. The mean heart rate at the fifth minute was significantly ($P < 0.05$) higher during active than during inactive and passive recovery.

Table 6.1 Cardiorespiratory variables before, during, and after dynamic knee extension and flexion exercise

	Rest	Exercise	Recovery
VO_2 (L/min)			
Inactive	0.23 ± 0.02	0.78 ± 0.06	0.22 ± 0.01
Passive	0.21 ± 0.03	0.77 ± 0.06	0.23 ± 0.01
Active	0.25 ± 0.01	0.84 ± 0.13	0.31 ± 0.02[a,b]
MBV (cm/s)			
Inactive	4.5 ± 0.6	13.1 ± 0.5	5.0 ± 0.6
Passive	4.2 ± 0.5	14.3 ± 0.7	7.7 ± 0.5[a]
Active	4.7 ± 0.5	15.8 ± 1.4	9.7 ± 1.1[a,b]
CO (L/min)			
Inactive	5.1 ± 0.4	9.3 ± 0.6	6.1 ± 0.3
Passive	5.3 ± 0.2	9.5 ± 0.5	6.5 ± 0.3
Active	5.3 ± 0.3	9.4 ± 0.4	7.2 ± 0.4[a,b]
HR (beats/min)			
Inactive	68.5 ± 5.1	98.7 ± 7.6	68.1 ± 5.6
Passive	67.2 ± 4.3	96.4 ± 6.3	67.8 ± 5.2
Active	64.4 ± 4.9	96.5 ± 7.5	73.5 ± 6.9[a,b]
SV (mL)			
Inactive	74.6 ± 5.4	94.4 ± 3.6	91.0 ± 6.6
Passive	80.1 ± 6.3	99.1 ± 3.8	96.8 ± 4.5
Active	82.7 ± 3.4	98.8 ± 6.3	99.9 ± 4.4
MBP (mmHg)			
Inactive	86.8 ± 3.1	100.7 ± 3.4	83.0 ± 3.4
Passive	88.0 ± 3.4	96.2 ± 3.4	82.6 ± 2.3
Active	84.6 ± 2.3	99.9 ± 4.7	85.5 ± 2.6
TPR ($mmHg \cdot L^{-1} \cdot min$)			
Inactive	17.9 ± 1.8	11.3 ± 1.0	13.9 ± 0.8
Passive	16.9 ± 1.2	10.5 ± 0.9	13.0 ± 0.8[a]
Active	16.5 ± 1.5	11.0 ± 0.9	12.1 ± 0.9[a,b]

Values are means ± SE for five subjects

Values for rest, exercise, and recovery were averaged at the second minute of rest, the fifth minute of exercise, and the fifth minute of recovery, respectively. *VO_2* oxygen uptake, *MBV* mean femoral artery blood velocity, *CO* cardiac output, *HR* heart rate, *SV* stroke volume, *MBP* mean arterial blood pressure, *TPR* total peripheral vascular resistance. Inactive, inactive recovery experiment; Passive, passive recovery experiment; Active, active recovery experiment

[a] $P < 0.05$, significantly different from inactive recovery

[b] $P < 0.05$, significantly different from passive recovery. Redrawn with permission from Takahashi et al.

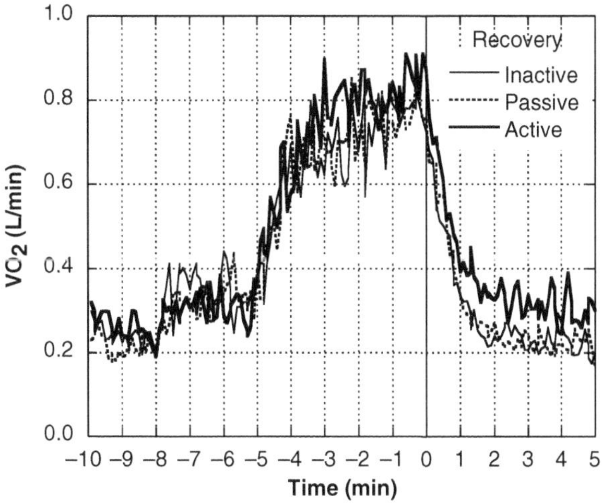

Fig. 6.3 Oxygen uptake (VO_2) during two-legged knee extension and flexion exercise at 60 W and recovery with no leg movement (inactive, thin line), passive leg movement (passive, dashed line), and active leg movement (active, thick line) at 0 W. Adapted with permission from Takahashi et al.

Figure 6.4b shows the changes in stroke volume during inactive, passive, and active recovery after the performance of the same rest and exercise. During the periods of rest, light exercise, and mild exercise, stroke volumes increased progressively to 80, 90, and 100 mL. After exercise, stroke volumes for three recovery modes all transiently increased somewhat from the preceding exercise levels and then decreased gradually; however, they remained above the resting levels during the 5 min of recovery.

Postexercise cardiac output in all recovery modes decreased gradually after an initial delay of about 20 s (Fig. 6.4c). The mean cardiac output at the fifth minute was significantly ($P < 0.05$) greater during active than during inactive and passive recovery; however, between inactive and passive recovery, the difference in mean cardiac output was insignificant. The greater cardiac output at the fifth minute during active recovery was mainly due to higher heart rate rather than higher stroke volume (Table 6.1).

Figure 6.5 shows the time courses of mean blood flow velocity in the femoral artery during inactive, passive, and active recovery. After exercise, all the blood flow velocities, regardless of recovery mode, initially overshot above the exercising levels, known as reactive hyperemia, and then gradually decreased separately in the descending order, active, passive, and inactive recovery. The peak values for mean blood velocity at 5 s after the end of exercise were 21.1 ± 2.4 cm/s for inactive, 23.7 ± 2.6 cm/s for passive, and 21.9 ± 1.7 cm/s for active recovery. The three peaks were significantly ($P < 0.05$) greater by about 54% than the steady-state levels (about 14.4 cm/s) during the antecedent exercise. The average of mean blood velocity at the fifth minute of recovery was significantly ($P < 0.05$) greater during active recovery, followed by passive recovery and then inactive recovery (Table 6.1).

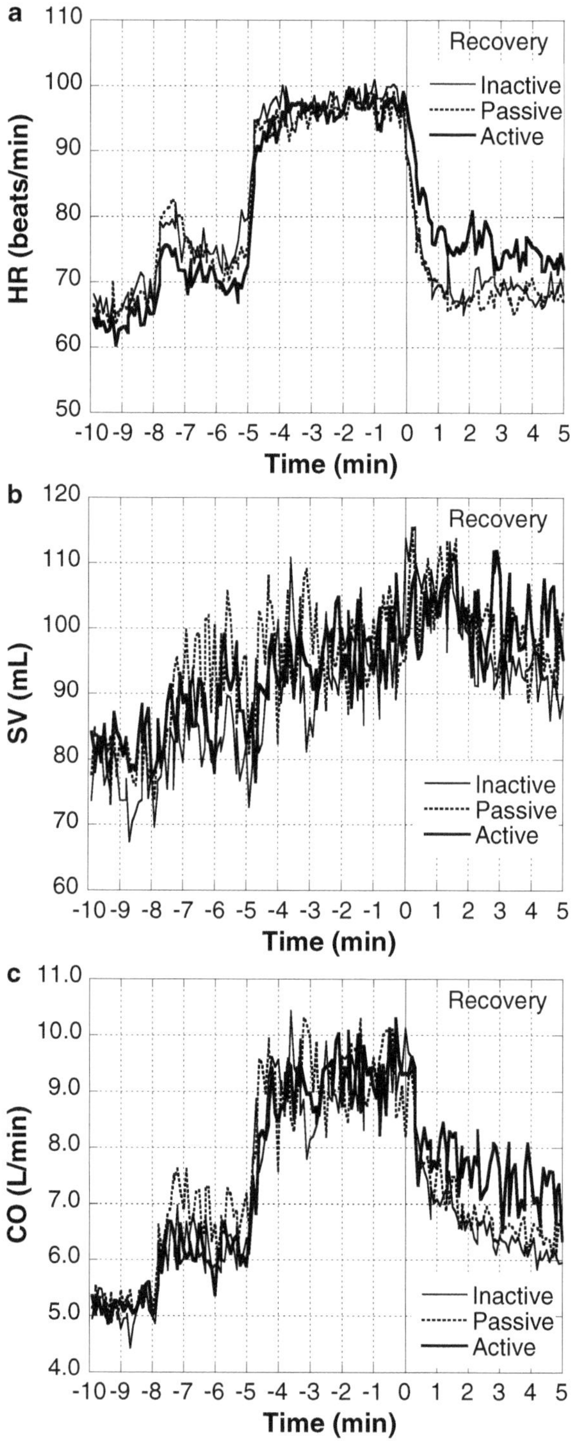

Fig. 6.4 Heart rate (HR, **a**), stroke volume (SV, **b**), and cardiac output (CO, **c**) during two-legged knee extension and flexion exercise at 60 W and recovery with no leg movement (inactive, thin line), passive leg movement (passive, dashed line), and active leg movement (active, thick line) at 0 W. Adapted with permission from Takahashi et al.

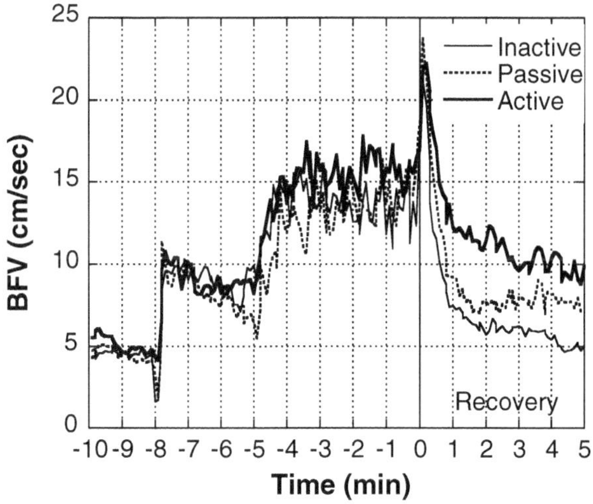

Fig. 6.5 Mean femoral artery blood velocity (MBV) during two-legged knee extension and flexion exercise at 60 W and recovery with no leg movement (inactive, thin line), passive leg movement (passive, dashed line), and active leg movement (active, thick line) at 0 W. Adapted with permission from Takahashi et al.

After cessation of exercise, mean arterial blood pressure decreased suddenly ($P < 0.05$) and remained relatively stable during inactive and passive recovery (Fig. 6.6b). Similarly, mean blood pressure during active recovery showed an initially rapid, significant ($P < 0.05$) decrease and a gradual return toward its pre-exercise resting level. As a result, there was no significant difference in mean blood pressure at the fifth minute between the three recovery modes.

Total peripheral vascular resistance showed an undershoot, consisting of an initially rapid decrease in the first 10 s and a rapid increase in the next 20 s ($P < 0.05$), followed by a gradual increase, regardless of recovery mode (Fig. 6.6c). The mean total peripheral resistance at the fifth minute was significantly ($P < 0.05$) lower during active recovery, followed by then passive and inactive recovery (Table 6.1).

The Doppler measurements of mean blood flow velocity at rest and during exercise at 0 and 60 W were highly reproducible in the same experiment repeated 1 week apart (Fig. 6.7). A linear relationship between the two measurements was represented by a Pearson coefficient of correlation as $r = 0.984$ ($P < 0.001$).

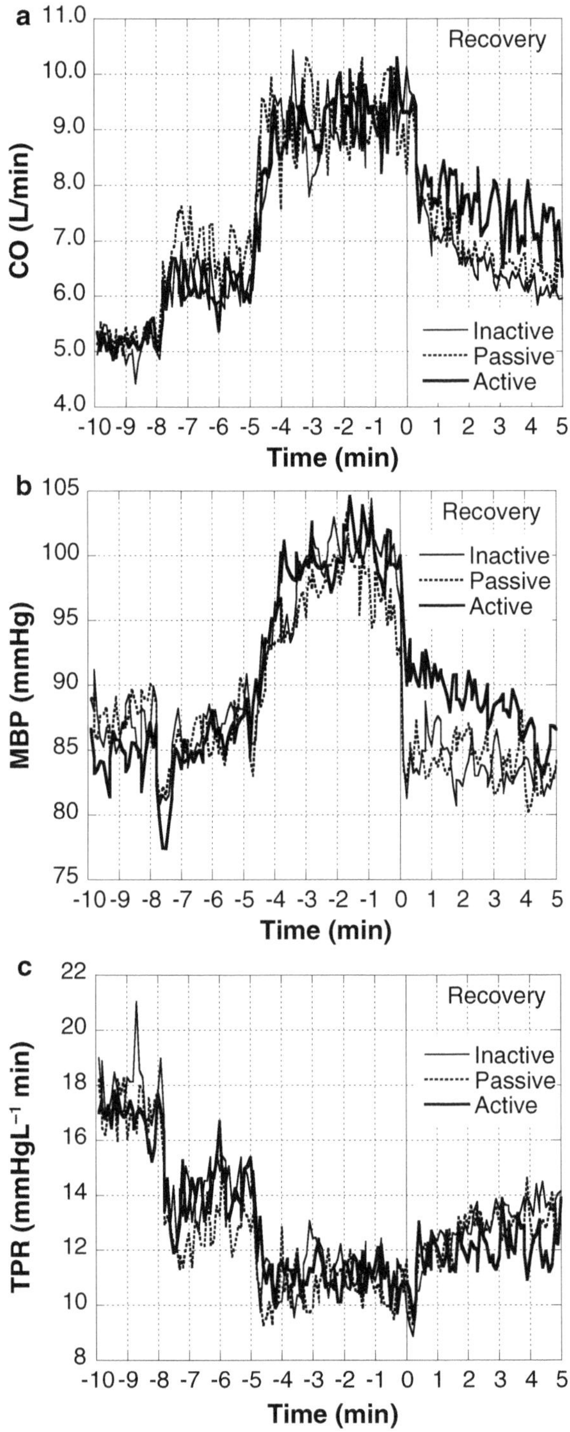

Fig. 6.6 Cardiac output (CO, **a**), mean arterial blood pressure (MBP, **b**), and total peripheral vascular resistance (TPR, **c**) during two-legged knee extension and flexion exercise at 60 W and recovery with no leg movement (inactive, thin line), passive leg movement (passive, dashed line), and active leg movement (active, thick line) at 0 W. Adapted with permission from Takahashi et al.

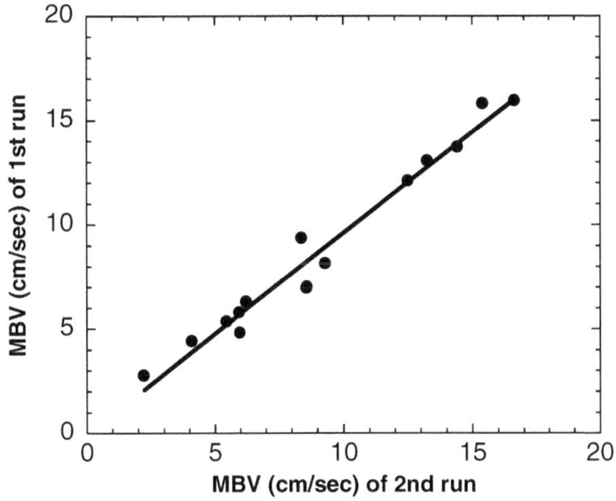

Fig. 6.7 Reproducibility of Doppler measurements of femoral artery blood velocity (MBV) at rest and during knee extension and flexion exercise both at 0 W and at 60 W. Data were obtained from five subjects who performed the same two runs at an interval of 1 week. The equation of regression line is expressed as: Y = –0.078 + 0.968X ($r = 0.984$, $n = 15$, $P < 0.001$). Adapted with permission from Takahashi et al.

6.4 Discussion

6.4.1 Blood Flow Velocity in Femoral Artery

After the end of exercise, mean blood flow velocity in the femoral artery showed a transient increase, reaching a peak within 5 s, followed by a rapid decrease within 1 min and then a gradual decrease in all three modes of the inactive, passive, and active recovery. The peak of mean blood velocity was almost equal in the three recovery modes. However, mean blood velocity at the fifth minute decreased to the individual levels in the descending order: active, passive, and inactive recovery. Conversely, total peripheral vascular resistance initially decreased temporarily and then slowly increased. The value of resistance at the fifth minute was lowest for active recovery, whereas that was highest for inactive recovery. The oxygen uptake at the fifth minute during active recovery was significantly greater than those during passive and inactive recovery, while the oxygen uptake at the fifth minute was not different between passive and inactive recovery. Further, the increased mean blood velocity in the femoral artery during passive recovery, compared with inactive recovery, was attributable to decreased total peripheral resistance without extra oxygen expenditure.

6.4.2 Mechanical Compression by Muscular Contraction

The initial rapid increase in the mean velocity of femoral artery blood flow after cessation of exercise, regardless of the recovery mode, appeared to be independent of changes in cardiac output and mean blood pressure; surprisingly, the initial increase in mean blood velocity occurred despite a sudden fall in mean blood pressure with unchanged cardiac output. Furthermore, the initial transient-decreased resistance and its subsequent gradual-increased total peripheral resistance were observed after the leg exercise of intensive muscular contractions was completed. These data suggest that blood flow to the exercising muscles was restricted at least in part by the mechanical compression of vasculatures due to intensive muscular contraction. This partially restricted blood flow was probably insufficient to meet the metabolic demands of the muscles during each contraction.

6.4.3 Non-consumption of Oxygen by Passive Muscle Contraction

If electromyogram measurements were made in this study, we could have confirmed that the subjects made no effort of their voluntary muscular contraction during passive recovery. In this study, the inactive and passive recovery modes had no significant difference in oxygen uptake. This result gives us indirect evidence that no induction of voluntary muscular contraction during passive recovery can be well accomplished with the assistance of the leg-pedal link system that we have developed. In addition, without extra oxygen expenditure for muscular contraction, mean blood velocity was greater during passive than during inactive recovery. On the other hand, it is well known that higher muscle blood flow during active recovery following exercise can enhance the restoration of the cardiovascular and respiratory systems through the removal or reuse of metabolic substances in active and inactive tissues and heat dispersion from active tissues by the circulation. Accordingly, the increased muscle blood flow induced by passive leg movement may also facilitate this process efficiently with very low energy expenditure or no extra oxygen consumption. In this respect, further studies should explore the physiological significance of passive recovery after intensive and prolonged exercise.

6.4.4 Variety of Measurement Methods

A value of 13–16 cm/s for mean blood velocity in the femoral artery during exercise in the present study is approximately one-half of the measurements reported in some studies. The differences in mean blood velocity between this study and other studies can be accounted for by several reasons. Firstly, different types of exercise, such as cycling, knee extension, and calf muscle contraction, put different mechanical stresses on the blood vessels in the muscle tissue. Secondly, these exercises vary in the size or amount of working muscle groups, which will affect mean blood velocity. Thirdly, as for the timing and duration of measurements, although our study continuously measured femoral artery blood velocity during contraction and relaxation of the muscles, other studies conducted the measurements between and after muscular contractions or during prolonged muscle relaxation. Fourthly, the site of the measurements at about 5- to 8-cm distal from the inguinal ligament is different than that at about 2-cm distal from the inguinal ligament. Finally, the mean blood velocity would be affected by the posture of subjects; for example, an upright sitting position vs. a supine position or a semi-supine position.

In preliminary experiments of this study on subjects in a sitting position, the Doppler velocimetry had failed to measure femoral artery blood velocity adequately at a distance of 2 cm from the inguinal ligament and before the bifurcation of the superficial and profound vessels. The movement of exercising legs was often a cause of signal artifacts that were included in the measurements. In the present study, the measurement probe was located at about 5 cm from the inguinal ligament and after the junction of the femoral artery to obtain clear Doppler spectrum profiles in the subjects in the upright sitting position. In contrast, in other studies, the measurement of blood flow velocity in the femoral artery in the proximity of the inguinal ligament (about 2 cm) should require the subjects to be in the supine or the semi-supine position during intensive leg exercise, although such a supine position is unnatural for usual physical activity except swimming.

6.4.5 *Validity of Blood Velocity Measurement*

What is important is that we should consider whether the measured values are reliable. The experimental set-up in this study enables us to perform the dynamic exercise of leg extension and flexion in the upright seated position with a cycle ergometer. In other studies, most of the experimental protocols have adopted different approaches to the problem of noise signals due to body motion. To prevent noise signals from being mixed into the measurements, they make subjects undergo intermittent static or concentric muscular contractions. In the present study, despite the difficulty of using the Doppler technique in measurements of blood flow in exercising muscles, the sensor of Doppler velocimeter remained relatively stable at the center of the target arterial vessel in sitting subjects who were performing the dynamic leg exercise. Furthermore, the relative change in mean blood velocity during inactive recovery above the resting baseline level in this study was similar to that observed in subjects following heel lifting in other studies. In addition, the averages of mean blood velocity at rest, during exercise, and during inactive recovery in this study are consistent with results reported from a study in subjects performing one-legged knee extension. Finally, as shown in Fig. 6.7, a correlation coefficient of 0.984 for mean blood velocity on measurements of separate days indicates that the Doppler-ultrasound velocimetry is a highly reproducible method for measuring dynamic changes in femoral artery blood velocity during dynamic knee extension and flexion exercise as well as at rest.

6.4.6 Shear Stress on Femoral Artery

The physical stress of blood flow on the blood vessel wall is a factor involved in control of blood flow and pressure. Blood flow to a vessel generates wall shear stress (WSS) on the inner surface of the vessel wall. The WSS (in dynes per centimeter), the flow-oriented stress acting as the viscous drag on the vascular endothelial surface, is given by the product of blood viscosity and shear rate of laminar blood flow. As blood flow is assumed to have constant viscosity, called Newtonian flow, through a cylindrical vessel, WSS is expressed as,

$$\text{WSS} = \mu \cdot \text{WSR} = \mu \frac{8V}{D} = 0.032 \times \frac{8 \times 15}{0.78} = 4.9$$

where WSR (=8*V*/*D*) is the wall shear rate, the mean blood flow velocity is $V =$ 15 cm/s (Table 6.1), superficial femoral artery $D = 0.78$ cm in diameter, and blood viscosity $\mu = 3.2$ cP (centipoise) = 0.032 P (poise) in reference to the literature. In this calculation, the vessel diameter and blood viscosity are assumed to be constant, since the diameter of the femoral artery before the bifurcation into the superficial and profound branches remains unchanged during cardiac cycles. Furthermore, according to Fåhraeus and Lindqvist, in blood vessels with an inner diameter of 1 mm or more, the viscosity of blood is constant at a normal value of 3.2 cP; for more information, see *Microcirculation of Fractal Branching Networks* written by Takahashi.

6.4.7 Reynolds Number

One characteristic of blood flow through relatively large blood vessels, such as arteries, can be defined as laminar or turbulent. The two types of flow can be classified by the Reynolds number ($Re = \rho VD/\mu$, dimensionless number). It states that $Re < 2000$ is of laminar flow and $Re > 2000$ is of turbulent flow. With the values of flow velocity (15 cm/s), vessel diameter (0.78 cm), and a blood density of $\rho = 1.00$ g/cm^3, the Reynolds number is estimated to be 366 less than 2000. This numerical value suggests that the femoral artery flow is in a laminar state during exercise.

6.4.8 Marry's Cube Law

The branching structure of vascular trees can be characterized by use of a mathematical model that links the relationship of a mother branch (D_0) and its daughter branches (D_1 and D_2) with a diameter exponent or a bifurcation exponent (m): namely, $D_0^m = D_1^m + D_2^m$. The exponent m has been theoretically determined to be 3.0

by the minimum work model of Murray and the minimum volume model of Kamiya and Togawa. It has been argued that the bifurcation exponent should be dependent on whether laminar flow ($m = 3.0$) or turbulent flow ($m = 2.33$) is present. The diameters of the common femoral artery and its superficial branch have been reported to be 1.05 and 0.78 cm, respectively, in the biomedical literature. In this case, an exponent of 2.332 can be obtained from the above equation on the assumption that the diameter of the profound branch is identical to that of the superficial branch.

6.4.9 Energy Loss at Bifurcation

The streamline of flow in the superficial branch at the junction of the common femoral artery may be disturbed, despite the lower Reynolds number suggesting a laminar flow state. This disturbance of femoral blood flow would be attributed to the flow pattern after the bifurcation, since the blood flow after a junction will require a certain distance for the reestablishment of the stability of laminar flow. The distance that is needed for the full development of laminar flow is referred to as the inlet length or entrance length (L), which is given by a mathematical expression, $L = 0.16\ D{\cdot}Re$. The inlet length is the length from a junction to a downstream point at which the deviation of flow velocity from the parabolic profile is less than 1%. This definition is valid when Re is greater than 50. The inlet length is the minimum distance needed for the full development of the laminar flow. An energy loss in a bifurcation region is greater than the loss occurring when laminar flow is fully developed. The transport efficiency of microcirculatory networks and the control of blood pressure and flow are reviewed in detail in Takahashi's textbook on the microcirculation.

6.4.10 Total Number of Capillaries in Lower Extremities

We shall now estimate the amount of skeletal muscle tissue that is exercising and the number of capillaries that are included in the tissue by means of a simulation model. According to the literature, the muscle mass of the whole body constitutes about 40% of the total body mass. The muscle mass of the lower body is also estimated to be about 70% of the total muscle mass. The average weight of the subjects participating in this study was 59.2 kg. Therewith for a 59.2 kg human, the muscle tissue mass of one-sided lower limb is calculated as follows: $59.2 \times 0.4 \times 0.7 \times 0.5 = 8.288$ kg.

Figure 6.8 schematically shows a component unit of muscle tissue surrounding a single capillary. This model is called the Krogh tissue cylinder model or simply Krogh cylinder model.

We consider the number of capillaries in one lower limb from the measured blood flow rate. The formulas that are used to calculate the number are as follows:

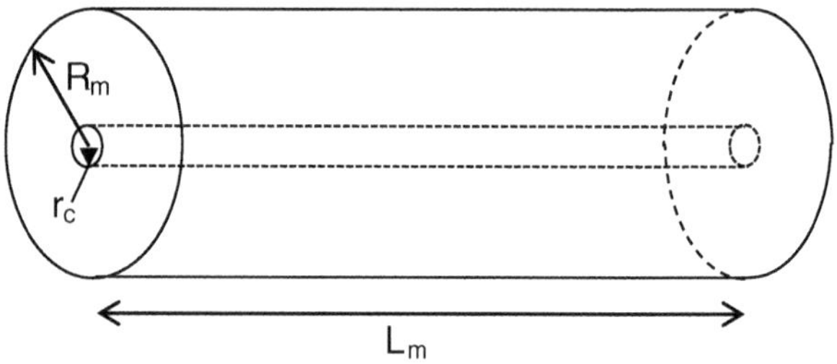

Fig. 6.8 Krogh tissue cylinder model. R_m the radius of the muscle tissue unit, r_c the capillary radius, L_m the length of the capillary and muscle tissue unit

$$F_m = \pi r_f^2 v_f \tag{6.1}$$

and

$$N_c = \frac{F_m}{\pi r_c^2 v_c} = \frac{r_f^2 v_f}{r_c^2 v_c} \tag{6.2}$$

where F_m is the blood flow of the common femoral artery, r_f the radius of the femoral artery, and v_f the velocity of blood flow in the femoral artery. From Eq. (6.2), the total number of capillaries (N_c) in the limb can be decided if the velocity of capillary flow (v_c) is given. According to the literature, v_c is given 1000 micrometers per second (μm/s). It is also assumed from the literature that the average of capillary radius r_c is 2.5 μm: specifically, 2.0 μm at the arterial end, and 3.0 μm at the venous end. In Eq. (6.1), F_m is determined to replace v_f with 14.4 cm/s as blood velocity measured during exercise in this study and r_f with 0.525 cm as the radius of common femoral artery. The result of the calculation is the number of capillaries, N_c = 2.5×10^9.

On the other hand, the volume (V_m) of unit of this tissue cylinder model is calculated as follows,

$$V_m = \pi R_m^2 L_m \tag{6.3}$$

$$M_m = \rho V_m N_m \tag{6.4}$$

$$N_m = N_c \tag{6.5}$$

and

$$R_m = \sqrt{\frac{M_m}{\rho \pi L_m N_c}} \tag{6.6}$$

where V_m is the volume of the muscle tissue unit including one capillary. R_m is the unit radius. The length of the unit (L_m) is an average of 700 μm from the literature. The muscle tissue mass (M_m) is 8.288 kg of one lower limb. The density (ρ) of muscle tissue is about 1.1 g/mL. N_m and N_c are the number of units and the number of capillaries, respectively, in the muscle tissues.

If N_c obtained from Eq. (6.2) is substituted for it in Eq. (6.6), the radius (R_m) of the muscle tissue unit will be 38.5 μm. If capillary blood flow velocity (v_c) is half of the value previously substituted for v_c in Eq. (6.2), that is, 500 μm/s, the numbers of capillaries (N_c) and units (N_m) are double those of the previous result. As a result, the radius (R_m) of the muscle tissue unit becomes 27.2 μm.

These results are derived from data under the conditions which were of 31% of the peak oxygen uptake as the relative intensity of the exercise. Thus, it is necessary to note that blood may not flow through all capillaries at the same time, but intermittent flow in each capillary may maintain the same tissue mass. At this time, the conservation law of blood flow must be observed. We also consider the number of capillaries at the maximum exercise intensity by using the same equations. Assuming that the cardiac output at the maximum exercise intensity is 30 L/min and then 85% of the cardiac output is distributed to both lower limbs, the blood flow of one limb is 148.75 mL/s (=30 × 0.85 × 0.7 × 0.5 L/min). Under these conditions, the number of capillaries is calculated as 7.6×10^9. In addition, the radius of the muscle tissue unit is 22.3 μm. Furthermore, 19.8 μm, subtracting a capillary radius of 2.5 μm from 22.3 μm, is the tissue distance that the capillary must nourish. This numerical value agrees well with a large number of literature values.

As mentioned above, at a relative exercise intensity of 31% of peak oxygen uptake, the number of capillaries was 2.5×10^9. This number is divided by the capillary number of 7.6×10^9 at the maximum exercise intensity, and the resulting number then will be 0.34, that is 34%. This numerical fraction suggests that capillaries are open to some extent at a rate proportional to the relative intensity of exercise.

It is important to consider the validity that the number of capillaries in the maximum exercise is 7.6×10^9. The number of capillaries will be contained in one lower limb. Now, let us move on to calculating the total number of capillaries in the muscle tissue of the whole body. The calculation results in $7.6 \times 10^9/0.7 \times 2 = 2.2 \times 10^{10}$. This result is almost in agreement with 1.7×10^{10} to 2.6×10^{10} of results reported by Dr. Kamiya and Dr. Folkow. For reference, a supplement explanation is added below. The size of muscle fiber changes depending on the type of oxygenation ability of muscles. For example, high-oxidative, slow-twitch fibers have small diameters, and low-oxidative, fast-twitch fibers have large diameters. The capillary number ranging from 1.7×10^{10} to 2.6×10^{10} has been estimated by Kamiya on the prudent assumption that the ratio of slow-twitch fibers to fast-twitch fibers is 20:80 in percent.

A mathematical model for microcirculation in fractal branching networks has been devised by Kamiya and Takahashi to offer new scope to study a scaling law which governs the number of the arterial ends of the capillaries in the vascular network. In that study, the number of the terminal arterioles (N_b) in the systemic circulation is expressed as a function of body mass (M_b, in kilograms) in a variety of mammals by the equation:

$$N_b = 2.202 \times 10^8 \, M_b^{\,1.04} \tag{6.7}$$

For a human of 59.2 kg body mass, the expected number of the arteriolar ends of the capillaries would be estimated as 153.47×10^8 from the above equation. Assuming that each of the terminal ends branches into four capillaries, the total number of the capillary vessels in the whole body can approximate 613.89×10^8. Since the muscle mass constitutes 40% of the body mass, the number of the capillaries in the skeletal muscle of the whole body can be estimated as 2.5×10^{10}. Surprisingly, the number is consonant with the numerical solution (2.2×10^{10}) of mathematical analysis of our model in this study. Incidentally, the 1.04, close to 1.00, power of body mass in Eq. (6.7) is of essential importance. This formula states that the density of the arteriolar termini and capillaries in the body is almost the same among all mammals ranging from a few grams to several tones.

6.5 Summary

The effect of the muscle pump on blood flow velocity in the femoral artery was examined in five healthy men during 5 min of recovery from upright sitting exercise of knee extension and flexion at 31% of peak oxygen uptake. The recovery phase was evaluated for cardiorespiratory variables in three different modes: (1) complete rest (inactive recovery); (2) passive-induced knee extension and flexion (passive recovery); and (3) voluntary knee extension and flexion (active recovery) at 12% of peak oxygen uptake. Blood velocity in the superficial femoral artery was measured beat-by-beat by Doppler-ultrasound velocimetry. After exercise, the mean blood flow velocity temporarily increased above the exercise level within 5 s followed by a rapid decrease and then a gradual decrease in all three recovery modes. However, the levels to which the mean blood velocity at the fifth minute returned were significantly different between the three recovery modes. The mean values of blood velocity were ranked in descending order for active, passive, and inactive recovery. The changes in total peripheral vascular resistance were opposite to the changes in mean blood velocity for each recovery; by definition, total peripheral resistance is calculated by dividing mean arterial blood pressure by cardiac output. Both cardiac output and oxygen uptake decreased rapidly during the first minute followed by gradual decreases, separately reaching either the unloaded exercise level during active recovery or the pre-exercise resting levels during inactive and passive recovery. The mean cardiac output and oxygen uptake at the fifth minute were significantly greater for active than for passive and inactive recovery. However, neither cardiac output nor oxygen uptake was distinctly different between passive and inactive recovery. These results demonstrate that during recovery from exercise, the increased mean blood velocity is maintained during passive knee extension and flexion without extra oxygen expenditure, as compared with inactive recovery. This phenomenon during passive recovery is chiefly attributable to decreased total peripheral resistance due to a weak effect of the muscle pump.

Further Reading

Takahashi T, Saitoh T, Okada A, Matsuo T. Differences in femoral artery blood velocity among active, inactive and passive recovery modes following knee extension and flexion exercise. Ther Res. 2006;27:1393–403.

Bangsbo J, Graham T, Johansen L, Saltin B. Muscle lactate metabolism in recovery from intense exhaustive exercise: impact of light exercise. J Appl Physiol. 1994;77:1890–5.

Bell HJ, Ramsaroop DM, Duffin J. The respiratory effects of two modes of passive exercise. Eur J Appl Physiol. 2003;88:544–52.

Shepherd JT. Importance of nitric oxide in muscular exercise. J Exerc Sci. 1996;6:1–10.

Boer P, Roos JC, Geyskes GG, Dorhout Mees EJ. Measurement of cardiac output by impedance cardiography under various conditions. Am J Phys. 1979;237:H491–6.

Charm SE, Kurland GS. Blood flow and microcirculation. New York: Wiley; 1974.

Fåhraeus R, Lindqvist T. The viscosity of the blood in narrow capillary tubes. Am J Phys. 1931;96:562–8.

Florian JA, Kosky JR, Ainslie K, Pang Z, Dull RO, Tarbell JM. Heparan sulfate proteoglycan is a mechanosensor on endothelial cells. Circ Res. 2003;93:e136–42.

Folkow B, Neil E. Circulation. Oxford: Oxford University Press; 1971.

Fung YC. Biomechanics: circulation. New York: Springer; 1984.

Hara N, Suzuki K, Mikami S, Uchida J, Seito N, Takahashi T. Combined therapy of extracorporeal shock waves and etidronate disodium as a potential treatment for post-traumatic myositis ossificans in the quadriceps muscle: a case report. J Physiother Phys Rehabil. 2020;5:1–5.

Hiatt WR, Yung Huang S, Regensteiner JG, Micco AJ, Ishimoto G, Manco-Johnson M, Drose J, Reeves JT. Venous occlusion plethysmography reduces arterial diameter and flow velocity. J Appl Physiol. 1989;66:2239–44.

Ishibazawa A, Nagaoka T, Takahashi T, Yamamoto K, Kamiya A, Ando J, Yoshida A. Effects of shear stress on the gene expressions of endothelial nitric oxide synthase, endothelin-1, and thrombomodulin in human retinal microvascular endothelial cells. Invest Ophthalmol Vis Sci. 2011;52:8496–504.

Joyner MJ, Casey DP. Regulation of increased blood flow (hyperemia) to muscles during exercise. Physiol Rev. 2015;95:549–601.

Kamiya A, Rippe B, Folkow B. Volume of the capillary exchange vessels measured with a gravimetric method in rat hindquarters. Microvasc Res. 1979;17:142–57.

Kamiya A, Takahashi T. Quantitative assessments of morphological and function properties of biological trees based on their fractal nature. J Appl Physiol. 2007;102:2315–23.

Kamiya A, Togawa T. Optimal branching structure of the vascular tree. Bull Math Biol. 1972;34:431–8.

Kamiya A, Togawa T. Adaptive regulation of wall shear stress to flow change in the canine artery. Am J Phys. 1980;239:H14–21.

Kamiya A, Togawa T, Yamamot A. Theoretical relationship between the optimal models of the vascular tree. Bull Math Biol. 1974;36:311–23.

Krogh A. Anatomie und physiologie der capillaren. Berlin: Springer; 1929.

Kubicek WG, Karnegis JM, Patterson RP, Witsoe DA, Matterson RH. Development and evaluation of an impedance cardiac output system. Aerosp Med. 1966;37:1208–12.

Laufs U, Urhausen A, Werner N, Scharhag J, Heitz A, Kissner G, Böhm M, Kindermann W, Nickenig G. Running exercise of different duration and intensity: effect on endothelial progenitor cells in healthy subjects. Eur J Cardiovasc Prev Rehabil. 2005;12:407–14.

Laughlin MH, Davis MJ, Secher NH, von Lieshout JJ, Arce-Esquivel AA, Simmons GH, Bender SB, Padilla J, Bache RJ, Merkus D, Duncker DJ. Peripheral circulation. Compr Physiol. 2012;2:321–447.

MacDonald MJ, Shoemaker JK, Tschakovsky ME, Hughson RL. Alveolar oxygen uptake and femoral artery blood flow dynamic in upright and supine leg exercise in humans. J Appl Physiol. 1998;85:1622–8.

Miyamoto Y, Higuchi J, Abe Y, Hiura T, Nakamura T, Mikami T. Dynamics of cardiac output and systolic time intervals in supine and upright exercise. J Appl Physiol. 1983;55:1674–81.
Murray CD. The physiological principle of minimum work. I. The vascular system and the cost blood volume. Proc Natl Acad Sci USA. 1926;12:207–14.
Pimsdale JE, Hartley LH, Guiney T, Ruskin JN, Greenblatt D. Post exercise peril: plasma catecholamines and exercise. JAMA J Am Med Assoc. 1984;251:630–2.
Pries AR, Neuhaus D, Gaehtgens P. Blood viscosity in tube flow: dependence on diameter and hematocrit. Am J Phys. 1992;263:H1770–8..
Rådegran G. Ultrasound Doppler estimates of femoral artery blood flow during dynamic knee extensor exercise in humans. J Appl Physiol. 1997;83:1383–8.
Schmidt-Nielsen K. Scaling: why is animal size so important? Cambridge: Cambridge University Press; 1984.
Shiotani I, Sato H, Yokayama H, Ohnishi Y, Hishida E, Kinjo K, Nakatani D, Kuzuya T, Hori M. Muscle pump-dependent self-perfusion mechanism in legs in normal subjects and patients with heart failure. J Appl Physiol. 2002;92:1647–54.
Shoemaker JK, Hodge L, Hughson RL. Cardiorespiratory kinetics and femoral artery blood velocity during dynamic knee extension exercise. J Appl Physiol. 1994;77:2625–32.
Takahashi T. Microcirculation in fractal branching networks. Tokyo: Springer; 2014.
Takahashi T. Easy to understand data analysis and statistics: let's start analysis statistics of medical and life science with excel. Tokyo: Ohmsha Ltd; 2017.
Takahashi T, Hayano J, Okada A, Saitoh T, Kamiya A. Effects of the muscle pump and body posture on cardiovascular responses during recovery from cycle exercise. Eur J Appl Physiol. 2005;94:576–83.
Takahashi T, Miyamoto Y. Influence of light physical activity on cardiac responses during recovery from exercise in humans. Eur J Appl Physiol. 1998;77:305–11.
Takahashi T, Nagaoka T, Yanagida H, Saitoh T, Kamiya A, Hein T, Kuo L, Yoshida A. A mathematical model for the distribution of hemodynamic parameters in the human retinal microvascular network. J Biorheol. 2009;23:77–86.
Takahashi T, Niizeki K, Miyamoto Y. Respiratory responses to passive and active recovery from exercise. Jpn J Physiol. 1997;47:59–65.
Takahashi T, Okada A, Hayano J, Tamura T. Influence of cool-down exercise on autonomic control of heart rate during recovery from dynamic exercise. Front Med Biol Eng. 2002;11:249–59.
Takahashi T, Okada A, Hayano J, Tamura T, Miyamoto Y. Influence of duration of cool-down exercise on recovery of heart rate in humans. Ther Res. 2000a;21:170–5.
Takahashi T, Okada A, Saitoh T, Hayano J, Miyamoto Y. Difference in human cardiovascular response between upright and supine recovery from upright cycle exercise. Eur J Appl Physiol. 2000b;81:233–9.
Takahashi T, Shibata M, Kamiya A. The optimum model of capillary-tissue arrangement in the skeletal muscle for oxygen transport to tissue (abstract). In: ISOTT2008 abstract book. The 36th annual meeting of International Society on oxygen transport to tissue. Tokyo: Japan Oxygen Society; 2008. p. 5.
Uylings HBM. Optimization of diameters and bifurcation angles in lung and vascular tree structures. Bull Math Biol. 1977;39:509–19.
Van Leeuwen BE, Barendsen GJ, Lubbers J, De Pater L. Calf blood flow and posture: Doppler ultrasound measurements during and after exercise. J Appl Physiol. 1992;72:1675–80.
Walløe L, Wesche J. Time course and magnitude of blood flow changes in the human quadriceps muscles during and following rhythmic exercise. J Physiol. 1988;405:257–73.
Wang JS, Jen CJ, Kung HC, Lin LJ, Hsiue TR, Chen HI. Different effects of strenuous exercise and moderate exercise on platelet function in men. Circulation. 1994;90:2877–85.
Whitmore RL. Rheology of the circulation. Oxford: Pergamon Press; 1968.

Chapter 7
Blood Flow Velocity in the Middle Cerebral Artery

7.1 Introduction

The brain of an adult person constitutes approximately only 2% of the body mass. Disproportionately, it receives roughly 15% of cardiac output and consumes nearly 20% of the total oxygen utilization at rest. Through the evolution of the human brain, the need for oxygen in brain tissue has been increasing with increasing cerebral tissue mass. In the mass of the brain relative to the body, a disproportionately large blood flow to the brain may be originally related to a delicate situation of or an adaptation of the cardiovascular system to the orthostatic stress in standing posture. Since the brain tissues are extremely fragile in oxygen deficiency, an adequate and continuous supply of oxygenated blood flow to the cerebral tissues is essential for sustaining the brain function as well as for keeping tissues alive. The blood flow in the middle cerebral artery (MCA) maintains about 80% of the flow volume arriving at the cerebral hemisphere. Changes in blood flow are also closely related to changes in cerebral metabolism.

Blood flow velocity in the middle cerebral artery has been reported to be significantly greater in women than in men. It has also been demonstrated that the diameter of the middle cerebral artery is 9% smaller in women than in men. In general, the calibers of the middle cerebral artery in both men and women are assumed to be constant within the normal range of systemic blood pressure. Thereupon an amount of oxygen transport via cerebral blood flow is proportional to changes in flow velocity that can be adjusted by perfusion pressure and vascular resistance.

Women have a low blood oxygen capacity, because the number of erythrocytes (red blood cells), which contain hemoglobin, is less in women than in men. The lower erythrocyte number (or lower hematocrit), which is accompanied with women's lower blood viscosity, is advantageous in decreasing the mechanical energy loss of blood flow against viscous resistances. Thus, the difference in the number of erythrocytes between women and men has one of decisive effects on the transport of

T. Takahashi, *The Cardiovascular System During Exercise and Recovery*,
https://doi.org/10.1007/978-981-99-8486-2_7

oxygen in the circulation. However, few attempts to date have been made to ascertain how the differences in hemological and hemorheological factors, such as blood oxygen capacity, hematocrit, and blood viscosity, between men and women, affect the regulation of blood velocity and flow volume in the cerebral circulation, both at rest and during exercise.

Therefore, this chapter is concerned with the differences in both blood velocity in the middle cerebral artery and perfusion pressure (the systemic artery blood pressure) between men and women at rest and during exercise; the blood velocity and arterial blood pressure are measured by transcranial Doppler ultrasoundgraphy and sphygmomanometry, respectively. It is also addressed that the physiological roles of hemological and hemorheological factors are implicated in variations in middle cerebral artery blood velocity in men and women at rest and during exercise. To clarify the effects of these factors on the regulation of blood velocity, we simulate blood flow with respect to oxygen transport through the middle cerebral artery by applying the physiological data available from the literature and the data obtained in this study, and then compare the results of numerical calculations between men and women.

7.2 Materials and Methods

7.2.1 Human Subjects

Of 20 young healthy volunteers, descriptive characteristics of 10 men and 10 women who participated in the present study are shown in Table 7.1. All subjects were fully informed about the procedures, risk, and benefits of this study. They gave written informed consent before the study. None of them had any history of cardiovascular, cerebrovascular, or respiratory disease. No one was engaged in any regular physical activity or involved in competitive sports. The study was approved by the University Institutional Review Board.

Table 7.1 Physical characteristics of the subjects

		Male (n = 10)	Female (n = 10)	
Age	Year (range)	21–24	21–24	NS
Weight	kg	60.7 ± 8.2	49.8 ± 4.4	$P < 0.05$
Height	m	1.69 ± 0.05	1.58 ± 0.04	$P < 0.05$
BMI	kg/m^2	21.1 ± 2.0	20.0 ± 1.7	NS
HR rest	beats/min	67.8 ± 8.3	78.0 ± 10.8	$P < 0.05$

Values are expressed as mean ± SD. *BMI* body mass index, *HR rest* resting heart rate

7.2.2 Experimental Design

The subjects were required to keep their body in an upright seated position on an electromagnetically braked cycle ergometer (EZ201, Combi Co. Ltd., Tokyo) during the period of rest throughout the exercise. They placed their feet on the footrest near the flywheel during the rest period. They rested for 15 min on the cycle ergometer and then exercised for 6 min. The work rate of cycle exercise was set constant at 100 W for men and at 50 W for women. The electromagnetically braked system of the ergometer was externally controlled to ensure the desired work rates. During exercise, the subjects pedaled at a constant rate of 60 revolutions/min paced by a metronome. The pedaling rate was monitored with an inductive revolution meter fitted to the ergometer.

Exercise tests were initiated at 10:00 for all subjects. No subjects were allowed to consume any beverages containing caffeine or alcohol after 21:00 the night before the exercising tests. They also agreed to desist from vigorous exercise for 36 h before the day of the tests. They had eaten a small breakfast at least 3 h before testing. They were familiarized with the test situation in several pilot experiments. All experiments were conducted at an ambient temperature of 21–22 °C.

7.2.3 Cardio- and Cerebrovascular Variables

The blood flow velocity in the right middle cerebral artery (MCA) was monitored with the use of a 2-megaheltz pulsed Doppler ultrasound system (Intra-view, Rimed, Tokyo). Its pulse-emitting probe was located over the temporal bone and the angle of insonation was adjusted for optimal Doppler signals. The probe attached to the skull at a fixed angle was held with a headgear (MD350FC, Rimed, Tokyo), which could adjust the position of probe, as shown in Fig. 7.1. The probe position was firmly secured in all subjects to the optimal insonation angle throughout the experiments. The blood flow velocity was assessed for 6–10 consecutive cardiac cycles recorded during the last minute at rest and during exercise. Systolic, diastolic, and mean blood flow velocity in the middle cerebral artery were denoted by *MCA_Vs*, *MCA_Vd*, and *MCA_Vm*, respectively.

The brachial artery blood pressure was measured by sphygmomanometry with a pneumatic arm cuff, which was held at heart level. Mean arterial blood pressure (MBP, in millimeters of mercury, mmHg) was calculated as diastolic blood pressure (DBP) plus one-third of pulse pressure. The blood pressure was recorded during the last minute at rest and during exercise.

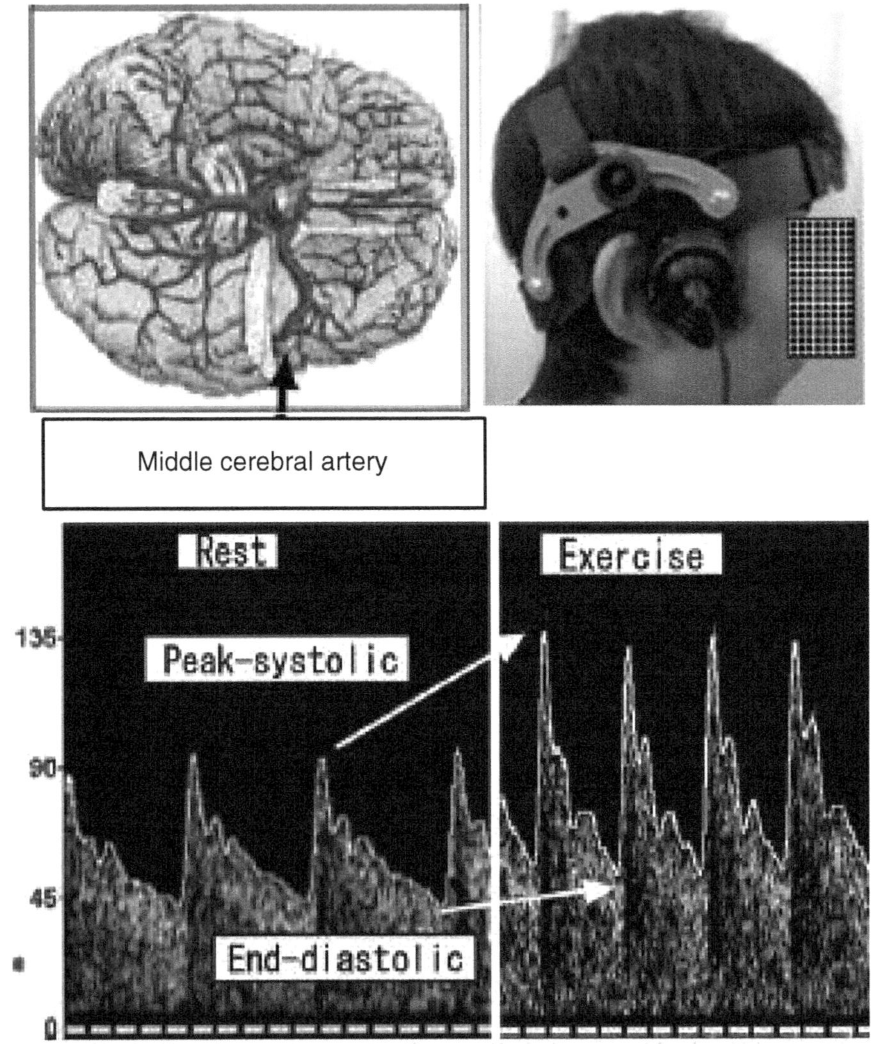

Fig. 7.1 The pulsed-Doppler probe was fixed onto the holder of a headgear to measure blood flow velocity in the middle cerebral artery (upper). The velocity profiles of blood flow measured in the middle cerebral artery during several cardiac cycles at rest (lower left) and during exercise (lower right). Although the angle of insonation is not determined precisely, the vessel is ideally at 0°. The errors of this angle uncertainty is minor, since most often, middle cerebral artery is insonated under an angle of less than 30°. If the angle of insonation ranges from 0 to 30°, cosine will vary between 1 and 0.87. Thus, the maximum error will be less than 14%

7.2.4 Statistical Analysis

Two-way analysis of variance was used to assess the effects of exercise and gender on mean values for all variables at rest and during exercise. When a significant *F* ratio was confirmed, Scheffé's post hoc test was used to identify significant differences. Differences were considered significant for all statistical analyses at $P < 0.05$. All values are expressed as mean ± standard deviation (SD).

7.3 Results

The workloads of cycle exercise were set at 100 W for 10 men and at 50 W for 10 women. Both the workloads were the stimuli to evoke similar cardiovascular responses in men and women, because there was no significant difference in heart rate between the two groups (119.8 ± 17.6 beats/min for men and 117.9 ± 10.3 beats/min for women).

Figure 7.2 shows the arterial blood pressure at rest and during exercise. There was no significant difference in systolic blood pressure (SBP) between men and

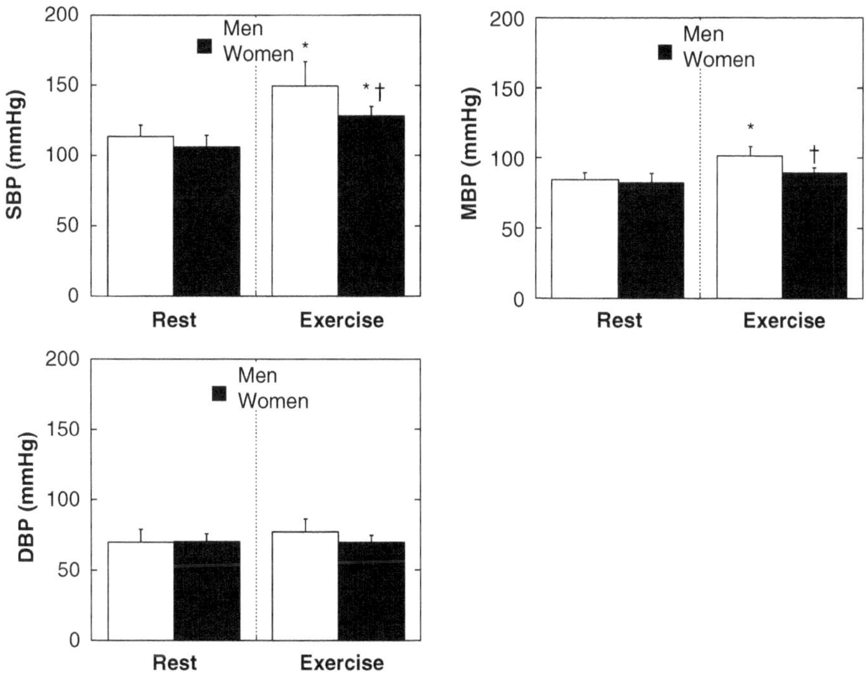

Fig. 7.2 Systolic arterial blood pressure (SBP, left), mean blood pressure (MBP, middle), and diastolic blood pressure (DBP, right) measured at rest and during exercise in men ($n = 10$) and women ($n = 10$) by sphygmomanometry. Values are means ± SD. *$P < 0.05$, significantly different from rest; †$P < 0.05$, significantly different from men. Adapted with permission from Watanabe et al

women at rest. The systolic blood pressure significantly (each $P < 0.05$) increased from rest to exercise in both men and women. The systolic blood pressure was significantly ($P < 0.05$) greater for men than for women during exercise. An increase in mean blood pressure (MBP) was significant ($P < 0.05$) for men from rest to exercise, whereas it was not statistically significant for women. There were no significant differences in diastolic blood pressure (DBP) between men and women, either at rest or during exercise. Diastolic blood pressure remained almost unchanged from rest to exercise in men and women.

Figure 7.3 shows the blood flow velocity in the middle cerebral artery at rest and during exercise. The systolic and mean blood flow velocity in the middle cerebral artery (MCA_Vs and MCA_Vm, respectively) in both men and women significantly (each $P < 0.05$) increased from rest to exercise. The averages of mean flow velocity in the middle cerebral artery for women were 63.5 ± 14.7 cm/s at rest and 72.8 ± 11.7 cm/s during exercise and for men were 48.6 ± 11.6 cm/s at rest and 57.5 ± 14.6 cm/s during exercise. Each value of the resting and exercising mean flow

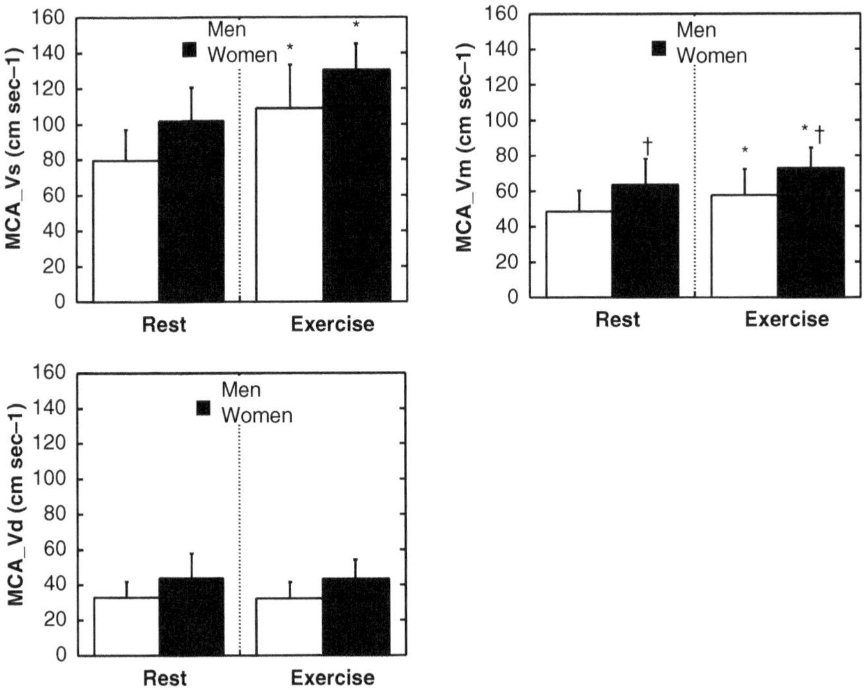

Fig. 7.3 Systolic blood flow velocity (MCA_Vs, left), mean blood flow velocity (MCA_Vm, middle), and diastolic blood flow velocity (MCA_Vd, right) in the middle cerebral artery measured by transcranial Doppler ultrasound velocimetry at rest and during exercise in men ($n = 10$) and women ($n = 10$). Values are means ± SD. *$P < 0.05$, significantly different from rest; †$P < 0.05$, significantly different from men. Adapted with permission from Watanabe et al

velocities was significantly ($P < 0.05$) greater for women than for men. The averages of systolic flow velocity in the middle cerebral artery at rest and during exercise were slightly but insignificantly greater for women than for men. Moreover, there were no significant differences in diastolic flow velocity in the middle cerebral artery either between men and women or between rest and exercise in either group.

Figure 7.4 shows the mean values for resistance index (RI), pulsatile ratio (PR), and pulsatility index (PI) in the middle cerebral circulation at rest and during exercise: by definition, RI = (systolic blood velocity – diastolic blood velocity)/systolic blood velocity, PR = systolic blood velocity/diastolic blood velocity, and PI = (systolic blood velocity – diastolic blood velocity)/mean blood velocity. The RI, PR, and PI increased significantly ($P < 0.05$) from rest to exercise in both men and women; however, there were no significant differences in PR, PI, or RI between men and women at rest or during exercise.

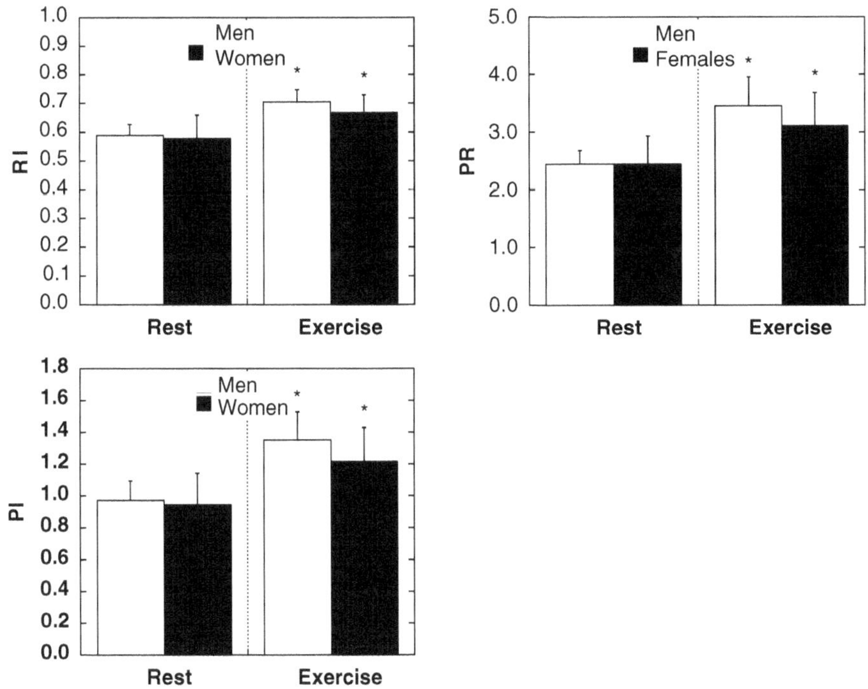

Fig. 7.4 Resistance index (RI, left), pulsatile ratio (PR, middle), pulsatility index (PI, right) with respect to blood flow velocity in the middle cerebral artery at rest and during exercise for men ($n = 10$) and women ($n = 10$). Values are means ± SD. *$P < 0.05$, significantly different from rest

7.4 Discussion

7.4.1 Blood Flow Velocity in Middle Cerebral Artery

The increased mean arterial blood pressure in response to exercise from the resting baseline value was significantly greater for men than for women in the upright seated position; however, the resting mean blood pressure was almost the same between the two groups. In contrast, the mean blood flow velocity in the middle cerebral artery (MCA_Vm) was higher in women than in men during rest and exercise. These results are in accordance with those of a study demonstrating that the mean flow velocity in the middle cerebral artery (MCA) for women was significantly greater than that for men at rest in supine posture.

7.4.2 High Blood Velocity for Oxygen Transport

The regulation of cerebral blood flow, especially the autoregulation, has been extensively studied in humans as well as in animals with respect to the influences of neurogenic, chemical, and physical factors. However, in this study, it is difficult to address additional information regarding autoregulation of cerebral blood flow in response to exercise. So we attempt here to explain the physiological significance of the greater flow velocity in the MCA for women by simulating the transport of oxygen (O_2) in the cerebral circulation and by comparing the results of simulation analysis between women and men. The numerical analysis is conducted with data obtained from the literature and this study, while a current simulation model is comprised of a set of formulas as follows:

$$\mu = 1.09 \times \exp(0.024 \times \mathrm{Hct}) \tag{7.1}$$

$$\mathrm{WSR} = \frac{4\mathrm{MCA\_Vm}}{R} \tag{7.2}$$

$$\mathrm{WSS} = 0.01 \times \mu \times \mathrm{WSR} \tag{7.3}$$

$$\mathrm{CaO_2} = \mathrm{SaO_2} / 100 \times 1.306 \times \mathrm{Hb} + 0.003 \times 95 \tag{7.4}$$

$$\mathrm{MCA\_Fm} = \pi R^2 \times \mathrm{MCA\_Vm} \tag{7.5}$$

$$O_2\,\mathrm{Transport} = \mathrm{MCA\_Fm} \times \mathrm{CaO_2} \tag{7.6}$$

and

$$\eta = \mathrm{O_2\,Transport} / \mu \tag{7.7}$$

where μ (cP) is the apparent blood viscosity at a shear rate of 1000 s^{-1} with respect to hematocrit (Hct, %), WSR is the wall shear rate (s^{-1}), WSS is the wall shear stress ($dyn{\cdot}cm^{-2}$), CaO_2 ($mL\ O_2{\cdot}mL^{-1}$ blood) is the oxygen content flowing to the MCA, SaO_2 (%) is the oxygen saturation in the MCA, Hb ($g{\cdot}dL^{-1}$ blood) is the hemoglobin concentration of blood, MCA_Fm ($mL{\cdot}min^{-1}$) is the mean volumetric flowrate, R (cm) is the radius of the MCA, O_2 *transport* ($mL\ O_2{\cdot}min^{-1}$) is the rate of delivery of oxygen through the MCA to the dependent brain tissues, and η ($mL\ O_2{\cdot}min^{-1}\ Poise^{-1}$) is the efficiency of oxygen transport through the MCA. The constant 1.306 in Eq. (7.4) is the oxygen capacity of hemoglobin in $mL{\cdot}g^{-1}$ Hb. The constant 0.003 in Eq. (7.4) is the solubility coefficient of oxygen to plasma per oxygen pressure in $ml{\cdot}dL^{-1}{\cdot}mmHg^{-1}$ and the oxygen partial pressure (PO_2) in the MCA is assumed to be 95 mmHg. The SaO_2 at the MCA is estimated to be 96.8% from Hill's equation with a half-saturation pressure (P_{50}) of 27 mmHg and an index of 2.7 for humans: $SaO_2 = [(PO_2/P_{50})^{2.7}]/[1 + (PO_2/P_{50})^{2.7}] \times 100$. The efficiency ($\eta$) of oxygen transport is evaluated by the ratio of oxygen transport to blood viscosity, the relation of which is analogous to the relationship between the optimum hematocrit and blood viscosity. The results of this simulation with the numerical parameters used for the calculation are summarized in Table 7.2. The mathematical and physiological meanings of this model and mathematical formulas are described in detail in Takahashi's textbook, *Microcirculation in Fractal Branching Networks*, so refer to them.

It has been reported in the literature that the caliber of the MCA, the blood hemoglobin concentration, and hematocrit are relatively low in women compared with men as listed in Table 7.2. Despite the narrower caliber of women's MCA, the

Table 7.2 Oxygen transport and physiological data used for simulation

		Men		Women		
Variables	Unit	Rest	Exercise	Rest	Exercise	Sources
MCA_Vm	cm s^{-1}	48.6	57.5	63.5	72.8	
R	cm	0.168	0.168	0.153	0.153	Müller
MCA_Fm	mL min^{-1}	257	304	278	319	Eq. (7.5)
Hct	%	45	45	42	42	de Simone
μ	cP	3.2	3.2	3.0	3.0	Eq. (7.1)
WSR	1 s^{-1}	1160	1374	1666	1909	Eq. (7.2)
WSS	dyn cm^{-2}	37.5	44.4	50.1	57.4	Eq. (7.3)
Hb	g dL^{-1} blood	15.8	15.8	13.9	13.9	Åstrand
O_2 content	mL O_2 mL^{-1} blood	0.203	0.203	0.179	0.179	Eq. (7.4)
O_2 transport	mL O_2 min^{-1}	52.0	61.6	49.7	57.0	Eq. (7.6)
η	mL O_2 min^{-1} $Poise^{-1}$	16.1	19.1	16.5	18.9	Eq. (7.7)
Re	Dimensionless	533	632	683	782	Folkow

MCA_Vm mean blood flow velocity in the middle cerebral artery (MCA), *R* radius of the MCA, *MCA_Fm* mean blood flow in the MCA, *Hct* hematocrit, *WSR* wall shear rate at the MCA, *WSS* wall shear stress at the MCA, *Hb* hemoglobin, η efficiency, *Re* Reynolds number; oxygen saturation at the MCA, 96.8%; oxygen tension at the MCA, 95 mmHg; Half saturation oxygen pressure, 27 mmHg. Adapted with permission from Watanabe et al volumetric flow rate is greater for women than for men at rest and during exercise, because of the significantly greater blood velocity in the MCA of women. The greater mean velocity of women is attributable in part to their lower hematocrit and lower blood viscosity. The reciprocal phenomena are consistent with the inverse relationships between cerebral blood flow and hematocrit and between cerebral blood flow and blood viscosity in other studies. Although the lower hematocrit is favorable for draining of the cerebral blood flow, the decreased hematocrit that leads to the reduced concentration of blood hemoglobin is accompanied by decreased blood oxygen content. Importantly, the increase in volumetric flow rate with the greater flow velocity in the MCA offsets the lower oxygen content of blood and the oxygen transport in women (49.7 mL $O_2 \cdot min^{-1}$ at rest and 57.0 mL $O_2 \cdot min^{-1}$ during exercise). Through this balance of these variables which complete one another, the oxygen transport in women becomes comparable to that in men (52.0 mL $O_2 \cdot min^{-1}$ at rest and 61.6 mL $O_2 \cdot min^{-1}$ during exercise). It is also important that the efficiency (η) of oxygen transport to the brain tissues is very similar between women and men under the resting and exercising conditions.

7.4.3 Shear Stress and Cerebrovascular Incidents

According to Eqs. (7.2) and (7.3), wall shear rate and wall shear stress are directly proportional to the blood flow velocity and inversely proportional to vessel radius. Wall shear rate and stress at the middle cerebral artery were considerably higher in women than in men at rest and during exercise, because of the greater flow velocity and the smaller radius for women. An estimated value of 37.5–50.1 $dyn \cdot cm^{-2}$ for wall shear stress at rest in both women and men are in agreement with the results obtained from arterial vessels of comparable size in our previous study on the vascular system based on its fractal nature. The relatively high wall shear stress at the middle cerebral artery for women with the lower hematocrit may play an important role in reducing the incidence of cerebral infarction and cerebrovascular incidents, including spasms and hemorrhages. In support of this argument, there are helpful reports demonstrating that the lower wall shear rate and stress inversely correlate with an increase in carotid intima-media thickness, reflecting arteriosclerosis. Remodeling of vascular walls has also been reported to be associated with the degree of wall shear stress to endothelial cells. However, the physiological implication of the gender difference in wall shear stress for the vessel wall is as yet unknown and needs further study.

7.4.4 Shear Stress on Common Carotid Artery

In contrast to the results of this study, wall shear stress on the common carotid artery has been reported to be significantly higher in men than in women; however, the difference declines with age. The higher wall shear stress in the common carotid artery for men than for women results not only from higher hematocrit (or viscosity), but also from higher flow velocity. The distinction in shear stress between the common carotid artery and the middle cerebral artery with respect to the gender difference in blood flow velocity remains unclear.

7.4.5 Flow Distribution of the Middle Cerebral Artery

In general, the motor-sensory regions related to the face, trunk, and hips (about 80% of the hemisphere) are supplied with blood flow through the middle cerebral artery; for supplemental information, the anterior cerebral artery supplies the motor-sensory area for the legs, and the posterior cerebral artery feeds on the primary visual cortex.

7.4.6 Effect of Body Temperature on Cerebral Artery Flow

When body temperature increases during dynamic exercise with large muscle groups, the associated elevation of brain metabolism may affect cerebral perfusion. Changes in brain temperature, however, cannot explain any regional change in flow velocity, since middle cerebral artery flow velocity decreases too rapidly to reflect changes in temperature of the brain after exercise.

7.4.7 Cerebral Blood Flow and Arterial Carbon Dioxide

Changes in the partial pressure of carbon dioxide (carbon dioxide tension) in arterial blood have a noticeable effect on cerebral blood flow. Figure 7.5 shows a relationship between blood velocity in the middle cerebral artery and end-tidal partial pressure of carbon dioxide in expired breaths during cycle exercise at a progressive intensity. As the exercise intensity ranges from 16 to 57% of maximal oxygen uptake, middle cerebral artery blood velocity is proportional to end-tidal partial pressure of carbon dioxide. Interestingly, during exercise continued above the anaerobic threshold (57% of maximal oxygen uptake), middle cerebral artery blood velocity decreases with decreasing end-tidal partial pressure of carbon dioxide. The later correlation between blood velocity and end-tidal partial pressure of carbon dioxide is moving in parallel to the former, thereby turning the anaerobic threshold in the opposite direction. The translation of the linear relationship between the two may be attributed to the difference in mean systemic arterial pressure.

Since the diameter of cerebrovascular vessels is greatly affected by changes in carbon dioxide tension in cerebral arterial blood, attention should be paid to the interpretation of blood flow that increases or decreases as a physiological response. In general, blood flow (f) in a vessel is defined mathematically as the product of the

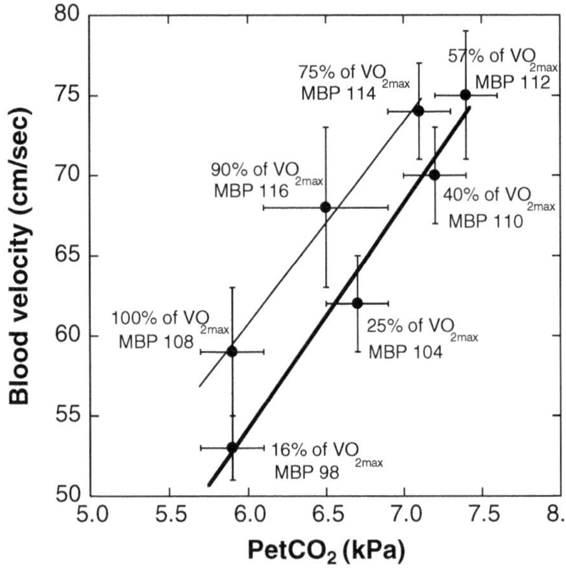

Fig. 7.5 Relationship between blood velocity in the middle cerebral artery and end-tidal partial pressure of carbon dioxide ($PetCO_2$) during cycle exercise. Blood velocity in the middle cerebral artery was measured with transcranial Doppler ultrasound velocimeter in 14 healthy young adults performing rest and dynamic exercise on a cycle ergometer at an intensity progressively increasing until exhaustion. VO_{2max} maximal oxygen uptake, *MBP* mean arterial blood pressure. Data from Moraine et al

Table 7.3 Percent changes in flow as a function of radius and velocity

Conditions	Velocity	Radius	Radius^2	Flow	% change of flow
1	1.00	1.00	1.00	3.14	1.00
2	1.10	1.00	1.00	3.46	1.10
3	1.00	1.10	1.21	3.80	1.21
4	0.90	1.10	1.21	3.42	1.09
5	1.10	0.90	0.81	2.80	0.89
6	1.10	0.95	0.91	3.13	1.00

Hagen-Poiseuille's law states that for a pressure drop (ΔP) between the inlet and outlet to a vessel segment with radius (r) and length (l) through which blood flows at volume flowrate (f) and mean blood velocity (v), their relation can be written in the forms, $f = \pi r^4 \Delta P/(8\ \mu l)$ and $v = r^2 \Delta P/(8\ \mu l)$

cross-sectional area of the vessel and the flow velocity ($\overline{v}$); here, the area is the square of the blood vessel radius multiplied by a constant, about $\pi = 3.14$, that is, $f = 3.14 r^2 \overline{v}$. Furthermore, if cerebral perfusion pressure (driving pressure, ΔP) is assumed to be constant, because it is relatively stable in the steady states during rest and exercise, theoretically the following approach can be adopted to the hemodynamics. Under these conditions, Poiseuille's law states the relationships of blood flow (f) to vessel radius (r) and velocity (v) in a vessel are expressed as $f = \pi r^4 \Delta P/(8\ \mu l)$ and $v = r^2 \Delta P/(8\ \mu l)$, where l is the length of the vessel. Hence, it is very important to understand the regular relationship of blood flow with vessel radius and flow velocity. For example, when in a vessel its radius increases by 10%, the effect of the radius contributes to 21% of changed blood flow, as shown under condition 3 in Table 7.3. Also, if flow velocity is reduced by 10% with a radius increase of 10%, the blood flow increases by 9% under condition 4 in Table 7.3. If flow velocity increases by 10% and vessel radius constricts by 5%, it is possible to have a condition in which we would observe no change in blood flow under condition 6 in Table 7.3. If neither flow velocity nor vessel radius changes, of course, there will be no change in blood flow.

Some studies have reported that the physical property of changes in vessel caliber is an important diagnostic tool in the interpretation of pathological states. Since blood vessels in the brain are sensitive to carbon dioxide, the calibers in reaction to carbon dioxide can be used as the standards for grading the vessels in response to other factors, such as body temperature.

7.4.8 Cerebral Blood Flow and Metabolisms

It is also important to note that the metabolic rates of glucose, oxygen, and lactate in the brain may be independent of the magnitude of changes in the cerebral blood flow during rest, exercise, and recovery. In this context, some authors have reported that although cerebral blood flow correlated with metabolic rate of oxygen in the resting state, regional uncoupling of cerebral blood flow and metabolic rates took place during neuronal activation induced by somatosensory stimulation. In such an experiment, flow-independent indexes, including the cerebral metabolic ratio (the cerebral molar uptake oxygen to glucose plus half lactate) and the arteriovenous difference of these components, are important for both measuring blood flow rate and studying brain function.

7.4.9 Limitations

In this study, the velocity of blood flow measured by transcranial Doppler velocimetry was used to assess changes in the flow rate in the middle cerebral artery at rest and during exercise. A previous study has demonstrated that relative changes in flow velocity in the middle cerebral artery measured by transcranial Doppler velocimetry correlate well with those in volumetric flow rate to the internal carotid artery measured by electromagnetic flowmetry in response to dynamic changes in systemic blood pressure. The flow velocity in the middle cerebral artery is directly proportional to the changed blood flow in the internal carotid artery, but is not accompanied with any substantial change in the cross-sectional area of the middle cerebral artery during moderate to heavy exercise. Furthermore, it has been reported that the wall of human cerebral arteries is very stiff in terms of its constituent elements exposed to mechanical forces. These results convincingly support the idea that the measurement of mean flow velocity is an effective means of estimating the blood flow in the middle cerebral artery even under conditions for normal changes in the perfusion pressure of the brain in response to light to moderate exercise.

Table 7.4 Distribution of cardiac output in a normal subject at rest and during light, heavy, and maximal exercise

	Rest	Light	Heavy	Max
VO_2 (mL/min)	250	700	2100	3500
Relative intensity (%)	7	20	60	100
Circulation (mL/min)				
Cardiac output	5000	11,400	21,000	30,000
Splanchnic	1200	1320	720	360
Renal	950	1080	720	300
Cerebral	650	900	900	900
Coronary	220	420	900	1200
Skeletal muscle	1030	5400	15,000	26,400
Skin	430	1800	2280	720
Other organs	520	480	480	120

Some of the data refers to data of Wade and Bishop and Folkow and Neil

From these facts, the present study assumed that the calibers of the MCA in men and women at rest and during exercise were constants in the calculations of hemodynamic and hemorheologic parameters. The validity of the vessel sizes used in our simulation is indirectly supported by the following estimations. The total of the blood flow in both the right and left middle cerebral arteries ($257 \times 2 = 514$ mL/min, Table 7.2) is comparable to the fraction of the systemic circulation ($5000 \times 0.13 \times 0.8 = 520$ mL/min, where $5000 \times 0.13 = 650$ mL/min for cerebral flow, see Table 7.4). The calculation formulas are as follows:

$$\text{Brain flow} = 13\% \text{of Cardiac output} \tag{7.8}$$

and

$$\text{Flow in middle cerebral artery} = 80\% \text{of Brain flow} \tag{7.9}$$

where cardiac output is 5.0 L/min in upright posture in Table 7.4.

A physiological feature of the brain receives special attention. It is the fact that blood flow velocity in the middle cerebral artery is markedly affected by changes in arterial carbon dioxide tension, as mentioned above. Moreover, it is not fully understood how the different phases of menstrual cycles affect the flow velocity on the one hand and the cardiorespiratory system on the other. Thus, further studies are necessary to elucidate these points.

7.5 Summary

The differences in arterial blood pressure and blood flow velocity in the middle cerebral artery between 10 young men and 10 young women were examined at rest and during a 6-minute bout of cycle exercise at a moderate intensity. The systolic, diastolic, and mean flow velocities in the middle cerebral artery were continuously measured by transcranial Doppler ultrasound velocimetry. The systolic, diastolic, and mean blood pressures were also recorded by sphygmomanometry. The mean flow velocity for men and women was significantly increased by 18.7% and 17.3%, respectively, from rest to exercise. The mean values for mean flow velocity at rest and during exercise were significantly greater in women than in men (rest, 63.5 ± 14.7 vs. 48.6 ± 11.6 cm/s; exercise, 72.8 ± 11.7 vs. 57.5 ± 14.6 cm/s). In contrast, the mean arterial blood pressure was significantly lower in women than in men during exercise, whereas that at rest was not significantly different between the two groups. Furthermore, a brain perfusion model was used to assess hemorheological significance under conditions of the greater flow velocity in the middle cerebral artery for women. The flowrate of blood through the middle cerebral artery under the circulatory regulation was important for delivering oxygen efficiently to the downstream tissues. The greater blood flow in the middle cerebral artery in women offset the lower arterial oxygen content, resulting in a similar amount of oxygen being transported by blood flow to that in men. These results suggest that the greater flow velocity in the middle cerebral artery in women, compensating for the lower arterial oxygen content, plays an important role in efficiently transporting oxygen through the bloodstream to tissues at levels comparable to those in men, both at rest and during exercise.

Further Reading

Watanabe S, Matsuo T, Sorimachi M, Yanagida H, Takahashi T. Comparison of blood flow velocity in the middle cerebral artery between men and women at rest and during exercise. Ther Res. 2009;30:537–44.

Severinghaus JW. Blood gas concentration. In: Fenn WO, Rahn H, editors. Handbook of physiology, vol. 2. Washington: American Physiological Society; 1965. p. 1475–87.

Birchard GF, Tenney SM. Relationship between blood-oxygen affinity and blood volume. Respir Physiol. 1990;83:365–74.

Moraine JJ, Lamotte M, Berré J, Niset G, Leduc A, Naeije R. Relationship of middle cerebral artery blood flow velocity to intensity during dynamic exercise in normal subjects. Eur J Appl Physiol. 1993;67:35–8.

Wade OL, Bishop JM. Cardiac output and regional blood flow. Oxford: Blackwell Scientific Publications; 1962.

Folkow B, Neil E. Circulation. London: Oxford University Press; 1971.

Aaslid R. Transcranial Doppler sonography. New York: Springer; 1986.

Aaslid R, Markwalder TM, Nornes H. Noninvasive transcranial Doppler ultrasound recording of flow velocity in basal cerebral arteries. J Neurosurg. 1982;57:769–74.

Ackerstaff RGA, Keunen RWM, van Pelt W, Montanban van Swijndregt AD, Stijnen T. Influence of biological factors on changes in mean cerebral blood flow velocity in normal ageing: a transcranial Doppler study. Neurol Res. 1990;12:187–91.

Åstrand PO, Rodahl K. Textbook of work physiology. New York: McGraw-Hill; 1986.

Chien S, Dormandy J, Ernst E, Mart A. Clinical hemorheology. Dordrecht: Martinus Nijhoff Publishers; 1987.

Chien S, Usami S, Skalak R. Blood flow in small tubes. In: Geiger SR, editor. Handbook of physiology, vol. 4. Bethesda: American Physiological Society; 1984. p. 217–49.

Fox PT, Raichle ME, Mintun MA, Dence C. Nonoxidative glucose consumption during focal physiologic neural activity. Science. 1988;241:462–4.
Fox PT, Raichle ME. Focal physiological uncoupling of cerebral blood flow and oxidative metabolism during somatosensory stimulation in human subjects. Proc Natl Acad Sci USA. 1986;83:1140–4.
Gnasso A, Carallo C, Irace C, Spagnuolo V, Novara GD, Mattioli PL, Pujia A. Association between intima-media thickness and wall shear stress in common carotid arteries in healthy male subjects. Circulation. 1996;94:3257–62.
Gregory IC. The oxygen and carbon monoxide capacities of foetal and adult blood. J Physiol. 1974;236:625–34.
Hayashi K, Handa H, Nagasawa S, Okumura A, Morita K. Stiffness and elastic behavior or human intracranial and extracranial arteries. J Biomech. 1980;13:175–84.
Heistad DD, Kontos HA. Cerebral circulation. In: Shepherd JT, Abboud FM, editors. Handbook of physiology, vol. 3. Bethesda: American Physiological Society; 1983. p. 137–82.
Hellström G, Fischer-Colbrie W, Wahlgren NG, Jogestrand T. Carotid artery blood flow and middle cerebral artery blood flow velocity during physical exercise. J Appl Physiol. 1996;81:413–8.
Humphrey PRD, Marshal J, Ross Russell RW, Wetherly-Mein G, Du Boulay GH, Pearson TC, Symon L, Zilkha E. Cerebral blood-flow and viscosity in relative polycythaemia. Lancet. 1979;314:873–7.
Ide K, Schmalbruch IK, Quistorff B, Horn A, Secher NH. Lactate, glucose and O_2 uptake in human brain during recovery from maximal exercise. J Physiol. 2000;522:159–64.
Ikegami M, Ichitani Y, Takahashi T, Iwasaki T. Copensatory increase in extracellular dopamine in the nucleus accumbens of adult rats with neonatal 6-hydroxydopamine treatment. Jpn J Neuropsychopharmacol. 2006;26:111–7.
Jacquez JA. Respiratory physiology. Washington: Hemisphere Publishing Corporation; 1979.
Jiang Y, Yanagida H, Takahashi T, Wang S. Bilateral frontal cortex activation during fragmented-letter identification is greater than that during complete-letter identification. Trans Jpn Soc Med Biol Eng. 2012;50:220–7.
Jørgensen LG. Transcranial Doppler ultrasound for cerebral perfusion. Acta Physiol Scand Suppl. 1995;625:1–44.
Kamiya A, Takahashi T. Quantitative assessments of morphological and functional properties of biological trees based on their fractal nature. J Appl Physiol. 2007;102:2315–23.
Kamiya A, Togawa T. Adaptive regulation of wall shear stress to flow changes in the canine carotid artery. Am J Phys. 1980;239:H14–21.
Kleinerman J, Sancetta SM. Effect of mild steady state exercise on cerebral and general hemodynamics of normal untrained subjects. J Clin Invest. 1955;34:945–6.
Kornet L, Lambregts J, Hoeks APG, Reneman RS. Differences in near-wall shear rate in the carotid artery within subjects are associated with different intima-media thicknesses. Arterioscler Thromb Vasc Biol. 1998;18:1877–84.
Krogh A. The comparative physiology of respiratory mechanisms. Philadelphia: University of Pennsylvania Press; 1941.
Laughlin MH, Davis MJ, Secher NH, von Lieshout JJ, Arce-Esquivel AA, Simmons GH, Bender SB, Padilla J, Bache RJ, Merkus D, Duncker DJ. Peripheral circulation. Compr Physiol. 2012;2:321–447.
Madsen PL, Hasselbalch SG, Hagemann LP, Olsen KS, Bülow J, Holm S, Wildschiødtz G, Paulson OB, Lassen NA. Persistent resetting of the cerebral oxygen/glucose uptake ratio by brain activation: evidence obtained with the Kety-Schmidt technique. J Cereb Blood Flow Metab. 1995;15:485–91.
Madsen PL, Sperling BK, Warming T, Schmidt JF, Secher NH, Wildschiødtz G, Holm S, Lassen NA. Middle cerebral artery blood velocity and cerebral blood flow and O_2 uptake during dynamic exercise. J Appl Physiol. 1993;74:245–50.

Markwalder TM, Grolimund P, Seiler RW, Roth F, Aaslid R. Dependency of blood flow velocity in the middle cerebral artery on end-tidal carbon dioxide partial pressure: a transcranial ultrasound Doppler study. J Cereb Blood Flow Metab. 1984;4:368–72.

Matsuo T, Watanabe S, Sorimachi M, Kanda M, Ohta Y, Takahashi T. Blood flow velocity waveforms in the middle cerebral artery at rest and during exercise. Artif Life Robot. 2011;16:398–402.

Matsuo T, Watanabe S, Sorimachi M, Takano H, Takahashi T. Blood flow velocity waveforms in the middle cerebral artery during cycle exercise and recovery. Artif Life Robot. 2015;20:336–40.

Müller HR, Brunhölzl C, Radü EW, Buser M. Sex and side differences of cerebral arterial caliber. Neuroradiology. 1991;33:212–6.

Newell DW, Aaslid R, Lam A, Mayberg TS, Winn HR. Comparison of flow velocity during dynamic autoregulation testing in humans. Stroke. 1994;25:793–7.

Niizeki K, Tokunaga S, Takahashi T, Miyamoto Y. Ventilatory responses to CO_2: constant fraction vs. constant inflow administration. Respir Physiol. 1996;104:91–106.

Quistorff B, Secher NH, Van Lieshout JJ. Lactate fuels the human brain during exercise. FASEB J. 2008;22:3443–9.

Samijo SK, Willigers JM, Barkhuysen R, Kitslaar PJEHM, Reneman RS, Brands PJ, Hoeks AGP. Wall shear stress in the human common carotid artery as a function of age and gender. Cardiovasc Res. 1998;39:515–22.

Schmetterer L, Findl O, Strenn K, Graselli U, Kastner J, Eichler HG, Wolzt M. Role of NO in the O_2 and CO_2 responsiveness of cerebral and ocular circulation in humans. Am J Phys. 1997;273:R2005–12.

de Simone G, Devereux RB, Chien S, Alderman MH, Atlas SA, Laragh JH. Relation of blood viscosity to demographic and physiologic variables and to cardiovascular risk factors in apparently normal adults. Circulation. 1990;81:107–17.

Smith JJ, Ebert TJ. Circulatory response to the upright posture. Boca Raton: CRC Press; 1990.

Takahashi T. Microcirculation in fractal branching networks. Tokyo: Springer; 2014.

Takahashi T. Easy to understand data analysis and statistics: let's start analysis statistics of medical and life science with excel. Tokyo: Ohmsha Ltd; 2017.

Takahashi T. The world of cognitive science: why can people read fragmented letters? Tokyo: Ohmsha Ltd; 2019.

Takahashi T, Jiang Y, Wang S, Ikegami M, Yoshimura A, Miyazawa R, Watanabe S, Matsuo T, Yanagida H. Identification of fragmented letters through minimum-distance interpolation. Ther Res. 2016;37:795–802.

Takahashi T, Niizeki K, Miyamoto Y. Respiratory responses to passive and active recovery from exercise. Jpn J Physiol. 1997;47:59–65.

Takahashi T, Niizeki K, Miyamoto Y. V_E-VCO_2 relationship in transient responses to step-load exercise from rest to recovery. Adv Exp Med Biol. 1998;450:207–12.

Takahashi T, Yamamoto S, Nishimura H, Watanabe S, Mimura Y, Saitoh T, Sugimoto T, Yanagida H, Yoshimura A. An allometric scaling mathematical model for capillday-covered alveoli in mammalian lungs. Ther Res. 2023;44:295–302.

Tohgi H, Yamanouchi H, Murakami M, Kameyama M. Importance of the hematocrit as a risk factor in cerebral infarction. Stroke. 1978;9:369–74.

Index

T. Takahashi, *The Cardiovascular System During Exercise and Recovery*,
https://doi.org/10.1007/978-981-99-8486-2

GPSR Compliance

The European Union's (EU) General Product Safety Regulation (GPSR) is a set of rules that requires consumer products to be safe and our obligations to ensure this.

If you have any concerns about our products, you can contact us on ProductSafety@springernature.com

In case Publisher is established outside the EU, the EU authorized representative is:

Springer Nature Customer Service Center GmbH
Europaplatz 3
69115 Heidelberg, Germany

Batch number: 09013545

Printed by Printforce, the Netherlands